A FISH CAUGHT IN TIME

Samantha Weinberg was born in London and educated at Trinity College, Cambridge. She has worked as a journalist in southern Africa, the United States and London, and is the author of *Last of the Pirates: the Search for Bob Denard*.

'A fascinating tale, mixing scientific intrigue, international politics and adventures on the high seas ... Samantha Weinberg makes the pathologically determined Smith – and his astoundingly devoted scientist wife, Margaret – the centre piece of her story. Although her tone is always sympathetic, [she] paints a picture of a man so driven that he even sleeps next to his coelacanth.' Gail Vines, *Independent*

'Many people have a vague recollection of the story of the coelacanth, but few of us appreciate the impact that this relic from the age of dinosaurs has had on our understanding of evolution. Fortunately, Samantha Weinberg has done an excellent job of reminding us how this ugly creature provided the missing link between life in water and life on land, spicing up her account by describing the daring exploits of the adventurers who went in search of the coelacanth.' Simon Singh, *Sunday Telegraph*

'Samantha Weinberg has cleverly made [the coelacanth] the subject of a well-researched, highly dramatised non-fiction narrative ... this book has all the ingredients of a bestseller: a curious, four-limbed fish to which we probably owe our own evolutionary existence, a host of eccentric ichthyologists, Teutonic submariners, British adventurers, and a series of political and environmental imbroglios.' Giles Foden, *Guardian*

'Part biography, part natural history and part evolutionary musing, *A Fish Caught in Time* comes garnished by great splashes of narrative colour . . . an excellent read.' Sara Wheeler, *Daily Telegraph*

'On a recent trip to South Africa my aunt was taken to see the original coelacanth and was not particularly impressed. What a shame that she hadn't the opportunity to read this entrancing book beforehand . . . Rest assured Ms Weinberg, I have already bought her a copy.' *Time Out*

'Fascinating . . . the book does not trawl cold scientific minutiae but presents science in the warm seas of adventure . . . The fish, although it cannot speak, ultimately says much about how humans view the world and their often grandiose self-configuration in it. A grand catch of a book.' *Irish Tribune*

'Passion and malice, glory and disappointment, greedy governments and gifted individuals all pack the pages of this truly thrilling tale. *A Fish Caught in Time* brings back the pure wonderment, the marvelling at the world we live in that we leave behind us with our school days.' *Independent* (South Africa)

'This fascinating study of an arcane subject and the scientists caught up in the adventure is full of incident and skulduggery. Weinberg holds us enthralled until the last page.' Sebastian Shakespeare, *Tatler*

'This is the best book I have read about the coelacanth. Samantha deals with everybody who has been connected with the coelacanth in a very live and human way. It took me back sixty years in a flash.' Dr Marjorie Courtenay-Latimer, finder of the 1938 coelacanth

A FISH CAUGHT IN TIME

The Search for the Coelacanth

Samantha Weinberg

FOURTH ESTATE • LONDON

To Mark
Hook, line and sinker

This paperback edition first published in 2000
First published in Great Britain in 1999 by
Fourth Estate Limited
6 Salem Road
London W2 4BU
www.4thestate.co.uk

Acknowledgements to Andre Deutsch Ltd for permission to
reproduce the poem 'The Coelacanth' by Ogden Nash (on page xv), taken from
Candy is Dandy: The Best of Ogden Nash.

1 3 5 7 9 10 8 6 4 2

A catalogue record for this book is available from the British Library.

ISBN 1-85702-907-0

Typeset in Monotype Garamond by Rowland Phototypesetting,
Bury St Edmunds, Suffolk
Printed in Great Britain by Clays Ltd, St Ives plc

CONTENTS

ILLUSTRATIONS

ACKNOWLEDGEMENTS

I saw my first coelacanth in the small museum in Moroni, the capital of the Comoros, in 1992. Before that, it had meant little to me – just a barely remembered word from long-past science lessons. Over the last few years, I have immersed myself in the coelacanth and its special world, but I could never have presumed to write a book about this most studied of creatures without the generous help and patience of coelacanth experts and junkies the world over.

There are several people without whom it literally would not have been possible. Robin Stobbs, latterly of the J. L. B. Smith Institute in Grahamstown, opened up his archives and his incredible personal store of knowledge and wisdom to me. He was a constant support and guide throughout the process, and remains a valued friend and e-mail buddy. To him I owe unlimited thanks. Marjorie Courtenay-Latimer endured thousands of my questions over several visits. Without her, the coelacanth would probably still not be known to the world, and I could certainly never have written about it. She is a truly special person, an inspiration to us all. Hans Fricke is an extraordinary adventurer, and knows more about the living coelacanth than anyone else – and he could not have been more generous in inviting me to see his submersible, *Geo*, in Germany, recounting his wonderful undersea voyages, and correcting my shaky science. Jerry Hamlin, creator of dinofish.com, is another true believer, devoted to saving the coelacanth. He gave me hours of his time. Thank you. I descended on the peaceful island paradise of Mark Erdmann and Arnaz Mehta with little notice, and for nine weeks lived next door and shared their boat, their dinners and their adventures. For eight months, while the discovery of the Indonesian population of coelacanths was still a secret, they trustingly kept me in the picture, and to them and to their team – Daeng Said, Tantes Ita and Meli – I dedicate the happy ending of this book.

The research process took a year, and encompassed four continents and numerous people. In the United States, thank you to Roy Caldwell and Keenan Smart, who shared the secret of the Indonesian coelacanth, and to Susan Jewett; to Eugene Balon in Canada; in England, to Anthony Gardner, who pushed me off in the right direction, to Henry van Moyland for his inspired title, to Peter Forey of the Natural History Museum and to Quentin Keynes, who visited the Comoros soon after the 1952 coelacanth adventure and kindly showed me his treasures; in Germany, to Karen Hissmann and Jürgen Schauer; in South Africa, to Jean Pote, Phil Heemstra, Paul Skelton and the entire staff of the J. L. B. Smith Institute of Ichthyology; to Bob and Gerd Smith, and to William Smith, who shared his memories of his illustrious and unorthodox parents and who generously gave me permission to quote from his father's book, *Old Fourlegs*, to Philip Tobias, Gill

Vernon of the East London Museum and Mike Bruton. Many old and new friends in the Comoros were a tremendous aid: Papa Claude, Christian Antoine and the staff of Le Galawa Beach hotel, who rescued us from deportation; Ali Toihir, Mouzaoir Abdallah, Mahmoud Aboud, the staff of CNDRS. A special thanks to Said Ahamada, who continues to work tirelessly for the Comoran coelacanth, and whose entire family welcomed me so warmly, and allowed me a glimpse into the world of the night fishermen; and to all the Comoran fishermen, who answered my questions with patience and humour. In Indonesia, thank you to Om Lameh Sonathon and his family, the most gracious of hosts, and to Om Maxon Haniko and their teams of fishermen, and Michael and Corrie and the ladies of MC.

Thank you to Gillon Aitken, agent and friend, and Emma Parry; to my editor, Virginia Bonham Carter, who has been with me every inch of the way; to James Kellow and the rest of the good folk at Fourth Estate. Thank you to my American editors, Larry Ashmead and Joe Wojak of HarperCollins, for their valuable input. Thank you to my family and my friends, who have been a constant support; my father and my sister, Joanna, the best of literary critics, my sister Kate, cousins Dan and Anne Simon, and my grandmother, Lilian Le Roith, whose car and house we took over during our three months in South Africa, and who remains my most uncritical supporter. Most of all, to Mark Fletcher, to whom this book is dedicated: the best travelling companion, editor and husband a girl could wish for.

Consider now the Coelacanth,
Our only living fossil,
Persistent as the amaranth,
And status quo apostle.
It jeers at fish unfossilized
As intellectual snobs elite;
Old Coelacanth, so unrevised
It doesn't know it's obsolete.

Ogden Nash

Southern Africa

I

LATIMERIA
CHALUMNAE

December in East London is hot and humid. An ochre haze smothers the small South African city; even the ocean breeze does little to dispel the seasonal lethargy. The year is 1938; *Gone With the Wind* is soon to open in America, and Hitler is menacing central Europe. But on the southern tip of Africa, three days before Christmas, most people's minds are on the approaching holidays: offices are beginning to close, families are drifting home to put the finishing touches to their festive arrangements.

At the East London Museum, the thoughts of the young curator, Marjorie Courtenay-Latimer, were far from the upcoming festivities. A small woman, with unruly dark hair and lively black eyes, she was surrounded by bones, racing to complete the assembly of a rare fossil dinosaur she and a friend had excavated in Tarkastad.

At quarter to ten in the morning, a shrill ringing echoed through the two small rooms of the tiny museum, shattering the young woman's concentration: the telephone had been installed only two days previously. Mr Jackson, manager of the Irvin & Johnson trawler fleet, informed her that Captain Hendrik Goosen had just arrived at the docks. 'There is a ton and a half of sharks for you on the trawler *Nerine*,' he said. 'Are you interested?' Marjorie was tempted to say no. She wanted desperately to complete the fossil display before the museum closed for the holidays, and she already had a load of fish specimens from Captain Goosen's last voyage waiting to be mounted. 'But I thought of how good everyone at Irvin & Johnson had been to me, and it being so near to Christmas, I thought the least I could do would be to go down to the docks to wish them the compliments of the season.' She grabbed a grain sack and called her native assistant, Enoch, and together they caught a taxi to the wharf.

'I went in to see Mr Jackson,' she recalls, sixty years later, 'and as I was going out, he said, "Well, I don't think it's quite a ton and a half of sharks, but a Happy Christmas to you!" They used to torment the life out of me.' She hitched up her cotton dress and climbed on to the 115-foot *Nerine*. The crew had all gone ashore except for an old Scotsman, who told her that the specimens were on the fo'c's'le deck. She looked at the pile of fish: sharks, seaweed, starfish, sponges, rat-tail fishes, all kinds of things. She told the Scotsman she probably would not be taking anything, nevertheless she sorted them out carefully. It was then that she noticed a blue fin sticking up from beneath the pile.

'I picked away the layers of slime to reveal the most beautiful fish I had ever seen,' she recounts. 'It was five foot long, a pale, mauvy blue with faint flecks of whitish spots; it had an iridescent silver-blue-green sheen all over. It was covered in hard scales, and it had four limb-like fins and a strange little puppy-dog tail. It was such a beautiful fish —

more like a big china ornament – but I didn't know what it was.'

'Yes, Miss, it's a strange one,' the old Scotsman said. 'I have been trawling for over thirty years, but I have never seen its like. It snapped at the captain's fingers as he looked at it in the trawl net. We thought you would be interested.' He told her that it had been trawled at a depth of forty fathoms, off the mouth of the Chalumna River, and that when Captain Goosen first saw it, he thought it so beautiful that he wanted to set it free. Marjorie said she would definitely take this one back to the museum.

She and Enoch eased the large fish – it weighed 127 pounds – into the sack and carried it to the taxi. The driver was horrified. 'I refuse to take any stinking fish in my new taxi!' he exclaimed. Marjorie replied, 'It is not stinking. It is perfectly fresh, and if that is the case, I will get another taxi. I brought you here to collect fish for the museum.' He relented and they carefully lowered the fish into the boot of the car.

'I was confused,' she relates. 'I had certainly never seen anything like it before, yet there was a voice nagging in my head. I kept on thinking back to school where I had written lines about a ganoid fish. I had a teacher, Sister Camilla, whose father was a palaeontologist at Uppsala University in Sweden, and he used to teach his daughter about marine palaeontology, and so she was always teaching us about fish. And on this day I wasn't paying attention, and she turned to me and said, "You-little-Latimer – what's a fossil fish?" And you-little-Latimer didn't know, because she hadn't been listening. "You-little-Latimer will write twenty-five lines: A ganoid fish is a fossil fish. A ganoid fish is a fossil fish." And you-little-Latimer wrote it out twenty-five times. I've still got the book. And so,

back at the museum, as I stared at the strange scales on the fish
– which indicated that it was a ganoid fish – those lines kept
going around in my head: A ganoid fish is a fossil fish – in
other words, a fish that has long since become extinct and is
known only from fossil records. The scales, the four limbs, all
pointed towards it being a ganoid fish. I was so near to
classifying it as a ganoid fish: but I thought it couldn't be a
fossil fish because it was alive. I didn't think it could be. But I
just knew it was something valuable.'

She searched through K. H. Barnard's *A Monograph on the
Marine Fishes of South Africa* and any other fish books she could
lay her hands on, but could find nothing resembling the
strange and beautiful specimen lying on the museum table.
That it was something unique and primitive, she could easily
tell by its curious structure, its head plates and fin formation.
She noted the extraordinary absence of blood or any slimy
discharge from its mouth, nose or body. She took measure-
ments and made rough sketches of the fish. At noon, the
chairman of the museum, Dr Bruce-Bays, dropped by and
Marjorie showed him her exciting new fish. 'He was a doctor
of medicine, a very sarcastic gentleman. He used to call me
Mistress Madge. "Mistress Madge, it's nothing but a rock
cod," he said. "You're making such a fuss about it, but it's
nothing but a rock cod. All your geese are swans."'

Many people would have given up at this point, and con-
signed the unidentified fish to the pre-Christmas rubbish
dump. But Marjorie was convinced she had something
special, and was determined to preserve the fish until she
could get someone to identify it. Enoch was sent to fetch a
small hand-cart, which a member of the board lent the
museum whenever they had to transport heavy items. When

he returned, they heaved the fish on to the cart and started to push their peculiar load into town.

'My first thought was to take it to the mortuary. At that time, the mortuary was right down at the park,' Marjorie recounts. 'So the two of us walked all the way there. It was very hot and everyone was very annoyed because they had to get out of the way for us. They shouted, saying why couldn't we walk on the street. Anyway, we got down to the hospital and I went to see the man in charge of the mortuary, a tall man by the name of Evans. When I suggested that our fish could go into his mortuary, he drew himself up to his full six feet, and with his eyes almost popping out of his head, he looked down his nose and said, "What an iniquitous request! Whatever is everybody going to say?" I said, "Agh, well they're all asleep, and it's such a beautiful fish." "No," he said. No, he couldn't possibly put a fish into the mortuary. Well, that was that.'

She next tried the cold storage. 'There was a gentleman there by the name of Latimer, no relation of ours. He at least had the decency to come and look at the fish, but he also said no, he wasn't putting any stinking fish into the cold storage. I don't suppose he was wrong, because it could have been giving off gases and there was food in the cold storage. So that was that with him, too.'

At the time, those were the only two refrigeration facilities large enough to accommodate a five-foot, nine-stone fish. Marjorie began to feel despondent: she knew she had to find some way of saving it. 'Then I thought of old Mr Robert Center, the taxidermist. He had known me since I was a little girl, and had taught me taxidermy, and I was sure he would help me somehow. By now, I realised more than ever that I had something very, very strange: those limb-like fins and the scales, all silvery and iridescent. It was so beautiful.'

They arrived at Mr Center's, and Marjorie showed him her fish and explained the difficulties she had experienced in trying to preserve it. She asked him whether he had ever seen a fish like it, and whether he had any idea what it could be. Mr Center looked at it; by then it was afternoon and the fish's colour was fading to a dusky grey. No, he admitted, he had never seen a fish like that and he had no idea at all what it could be. He told Marjorie to put it on the table in his workroom. 'If we could get some formalin, we could wrap it up and perhaps get someone from Rhodes University to identify it,' he suggested. Marjorie agreed. 'I said yes, I will get Dr Smith [as he was then].'

She went to a chemist friend in search of formalin with which to preserve the fish. It was at that time scarce, and only a few chemists kept stock for the hospital, so she was only able to get a little – about a litre – which she took back to Center. 'We diluted it, and we got *Daily Dispatch* papers, saturated them in the liquid and then wrapped the fish up, very carefully. We needed a sheet of some sort. We asked Mrs Center, but she wasn't parting with a sheet, so I walked all the way home – there were no buses or anything – and I explained to Mommy what was happening and that I needed to wrap the fish in some cloth so it wouldn't go off until I could get in touch with Dr Smith. She gave me a double bed sheet. We took it and wrapped the fish really beautifully, with all the formalin-soaked *Daily Dispatch* pages inside.'

Dr James Leonard Brierley Smith, a chemistry lecturer at Rhodes University in Grahamstown and amateur ichthyologist, acted as honorary curator of fishes for the smaller museums along the south coast. Marjorie had first met him five years previously, on the coast, when she was busy collecting shells and unusual seaweeds for the museum. A spry man, almost drowned by his baggy shorts, with startling blue eyes and a thick bristle of sandy hair, had come up to her and asked what she was doing; she had explained that she worked at East London Museum, and so the friendship began. Smith often visited Marjorie at the museum and helped her to classify unusual fishes. He was not, however, at Rhodes when she tried to telephone him on 22 December 1938. She left a message, but when he hadn't got back to her by the next day, she wrote to him, enclosing a rough sketch of the fish.

Daily, she waited for his reply. 'Nothing happened, nothing happened, nothing happened,' she recalls. Christmas passed in a fishy blur. 'I began to feel very despondent. I couldn't think about anything but the fish. My family couldn't understand what was the matter, but I just knew, deep in my bones, that it was something important.' Then Boxing Day: still no word. 'I checked the post every day and waited for a phone call, but there was no word from Smith.'

It was a shimmering hot week. Every afternoon Marjorie used to go down to Mr Center to check on the fish and although it all looked intact, by the 27th, oil had started seeping out. Center said he was worried that the loss of oil would begin to destroy the fish. Marjorie didn't want to risk it, so she gave him the go-ahead to skin it, only not from the side as was normal practice in mounting fish in those days, but rather right down underneath the stomach, so as to save all the scales. It was a hard, slow job. Center cut carefully through the thick scales. The flesh below was pure white and could be

EAST LONDON MUSEUM

ALL SPECIMENS AND EXHIBITS FOR
THE MUSEUM TRAVEL FREE BY POST
OR RAIL IF ADDRESSED:
O.H.M.S.
CURATOR. MUSEUM. EAST LONDON.
PHONE 2995.

East London
SOUTH AFRICA.

23ʳᵈ *Dec.* 193 8.

Dear Dr Smith,

I had the most queer looking specimen brought to notice
yesterday, The Capt of The trawler told me about it so I immediately
set off to see the specimen which I had removed to aur Taxidermist
as soon as I could . I however have drawn a very rough sketch and
am in hopes that you may be able to assist me in classing it.

It was trawled off Chulmna coast at about 40 fathoms.

It is coated in heavy scales, almost armour like., the
fins resemble limbs, and are scaled right up to a fringe of filment.

The Spinous dorsal, has tiny white spines down each filment.

Note drawing inked in in red.

I would be so pleased if you could let me Know
what you think , though I know just how difficult it is from a
discription of this kind.

Wishing you all happiness for the season.

Yours Sincerely.

M. Courtenay-Latimer

worked like clay. It did not appear fibrous, nor like any fish
flesh either of them had seen before. There were no ribs and
there was only a flexible tube where the spine should have
been. As Center cut into it, pale yellow oil just poured out.
Marjorie saved a whole bottle of the fine oil, which she set

aside for J. L. B. Smith, and the hard, bony tongue, which she took home to examine. She told Mr Center that if she had not heard from Dr Smith by the next evening, he could dump the insides and continue with the mounting.

'There was no word. We waited and waited every day for a letter from Dr Smith. It was terrible, really. The New Year passed, and we were still waiting.' It would be an excruciating thirteen days before Marjorie eventually heard from J. L. B. Smith.

Although Marjorie Courtenay-Latimer was not a trained ichthyologist, she was certainly knowledgeable in the field. Even before she was born, she was destined to become a naturalist. Her father, Eric Henry Courtenay-Latimer, wrote in his diary two months before her birth, '... Willie [his wife] and I both pray it will be a lover of all that is beautiful in nature. Willie wants it to be a botanist – I want it to be a lover of animals and birds.'

Their first daughter, Marjorie Eileen Doris Courtenay-Latimer, was born more than two months prematurely on 24 February 1907 (a Piscean, appropriately). She weighed a mere one and a half pounds and was given little chance of survival. To her parents, she was a thing of wonder, and they recorded every detail of her life. 'The tiny morsel looked like a miniature doll wrapped in cotton wool,' her father wrote. 'It had masses of dark hair, no finger-nails or toe-nails ... Our little treasure is the most fascinating baby in East London ... our joy in [her] seems more so that it is frail and small.'

The Courtenay-Latimers were not a wealthy family, and they had a fairly itinerant lifestyle, following Eric Courtenay-Latimer's job in the South African Railways from one station to the next. Yet they were happy, and took enormous pleasure in outdoor life, in picnics and long seaside walks. On her first birthday, they took their still fragile daughter to Cape Agulhas, on the southernmost tip of the African continent: 'Margie was thrilled with the beach and fell in love with a special shell which she played with all day, and finally fell asleep hanging on to it,' her father recorded. Aged two, she was found in the duck pond – she had collected all her aunt's ducklings in her pinafore and wanted to take them to bed with her.

Despite her healthy lifestyle, she was often ill. On several occasions, it looked as if she was going to die. 'She is thin and frail,' her father wrote, 'but has a very determined little character, a quaint, little, serious-minded child, with a deep affection for animals, birds, flowers, and her mother and sister.'

Birds were Marjorie's special interest: she used to spend long hours watching their nests, collecting eggs and feathers, studying their behaviour. When she was eleven, she announced that she would write a book on birds some day. She also had a wonderful butterfly collection, and started gathering ferns and old stone implements. She was a great defender of animal rights, once getting into a fight with her cousins when they wanted to drop a kitten down a well. She came into conflict with her parents, one day, over a lily: 'Margie got a whipping and was sent to bed because she argued with her elders about a Lotus Lily,' her father wrote. 'She insists it is a Water Lily and not a Lotus. When I asked her why, she said, "Lotus is a dreadful name for a beautiful plant and Water Lily is a beautiful name."'

Margie excelled at school and came top of her class in

everything except mathematics. Her father wished he could afford to send her to boarding school, but Dr Brownlee, one of the many botanists Margie had befriended, reassured him: 'No teaching could be found to teach her what she loves best in life; the beauty of nature is a gift that she has inherited – the knowledge she has gained in her short life is not found within the four walls of a school. Latimer, be patient, this child will go far in her natural gift of the beautiful. God grant her the health.' When Margie nearly died of diphtheria, Dr Brownlee stayed with her every night and nursed her back to health.

At fifteen, she was sent to a convent school, where she first came into contact with Sister Camilla and her fossil fishes. She continued to do well in all subjects, including music: 'She has a lovely touch,' her father recorded. 'In a way she seems to speak her nature in her playing. She has grown a very gentle loving girl, always ready to help others, and is a great help to her mother. Not pretty, but has a refined face with sparkling, mischievous eyes when she laughs.'

Now in her early nineties, Marjorie Courtenay-Latimer still has sparkling eyes when she laughs, which is often. She lives in East London with Cindy, her fox terrier, in a small house next door to the one she used to live in with her family – where she spent the agonising Christmas of 1938. There are books everywhere – mainly about nature – and jars of shells, woven baskets, flowers and a life-sized, not-yet-finished, red clay model of a Xhosa woman. Another unfinished sculpture is covered by cloths; she lifts them to reveal a startling like-ness of J. L. B. Smith. She has recently taken up painting flowers on ceramic tiles, which are placed carefully to dry on the window-sill. Almost seventy years have passed since she started work at the East London Museum, but her time there, first as curator and then as director, was clearly the most important period in her life.

'I always dreamed of working in a museum,' she recounts. 'Second to that, I wanted to be a nurse.' When she was twenty-one, she became engaged to Alfred Hill, a childhood friend. He was a handsome, party-loving boy. 'We used to go up and canoodle on the hill near my mother's people's farm on Addo Heights. From there we could see the lights of the Bird Island lighthouse sweeping across the sea.' Marjorie developed a compulsive desire to visit the tiny, remote isle. A year later, she broke off her engagement: 'He didn't like my madness in collecting plants and climbing trees after birds,' she recalls with a chuckle. 'He said it was a sissy game and no wife of his was going to do that. Then I fell in love with Eric Wilson, whose father owned the steel factory, but he died and I was heartbroken. He was the love of my life and I never fell in love again.'

She decided to become a nurse and got a place on a training course in King William's Town. A few weeks before she was to have started, however, she was invited by one of her naturalist friends, Dr George Rattree, to chase her dream and apply for the job of curator of East London's first museum – then still under construction. 'I went to meet the board, which consisted of the mayor and all these old gentlemen. I was shivering with fright. Then Dr Bruce-Bays (the chairman) asked me if I played the piano. "Yes," I said, so low and weak I wondered if he heard me. They asked me all sorts of questions about what I liked doing. Dr Rattree said, "Do you know anything about platanna?" Now a platanna is a frog, and I said, "Oh yes," and told him chapter and verse about the breeding of these things and where they could and couldn't be found. Twenty-five girls were interviewed that morning, and they were all beautifully dressed. I was wearing a home-made dress with bluebells of Scotland all over it and a funny little straw hat. I didn't think I would get the billet.' Nine days later, she was offered the job at a salary of £2 per month. She was to be in sole charge of setting up the displays of specimens and running the museum. It was her first job. She was twenty-four.

'I took over the museum in August 1931. It was an empty shell. They had a little room downstairs where they had a terrific lot of rubbish. There were six birds, which were riddled with demestes [a flesh-eating beetle], so I burnt them — it's a wonder they didn't sack me on the spot. They had a bottled piglet with six legs, about twelve pictures of East London which were quite nice, and twelve prints of Kaffir War scenes. That was it. And there was a box of stone implements collected by Dr Bruce-Bays, which were no more stone implements than my foot was a stone implement. They all went into the dump.'

She went home on her first day, consumed with doubts as to how she was going to be able to fill the museum. The next day she returned with an axe and destroyed the 'terrible, terrible' display cases, that had been built with a donation from a local philanthropist. She gathered together some old evening dresses, china and jewellery, and stone implements from her own collection, which, with her mother's collection of beadwork dating right back to 1858, and her great-aunt Lavinia Walton's dodo egg, she made the museum display.

From the day the museum opened, it became Marjorie's life. 'I used to go out on weekends and collect wild flowers which I would label and try and teach the children their names.' She never stopped collecting; every weekend, every holiday, adding large collections of South African sea shells, seaweeds, birds' eggs, butterflies, moths, insects and local ethnological material to the little museum's growing display.

East London Museum earned a good reputation. In 1932, two visiting dignitaries were so impressed by 'the lady curator' (as Marjorie refers to herself) that they organised for her to spend six months at the Durban Museum. There she learnt how to mount, model and classify all manner of specimens. She returned to her own museum full of enthusiasm and ambition.

In December 1933, Marjorie met J. L. B. Smith for the first time, and he encouraged her to send him any fish she needed classified. 'I was very fond of Professor Smith,' she says. 'People thought he was a very difficult man, but I always got on with difficult people. He was very exacting and I was scared of getting things wrong in front of him. But I loved and admired old JLB. He was the most wonderful person. I was very lucky to have such wonderful friends.'

In 1933, she was invited to spend six months at the South African Museum in Cape Town. While there, she met a Mr Patterson who was in charge of all the offshore islands, including Bird Island, which she used to gaze at as she canoodled with Alfred Hill. 'I have always believed that I had callings in life: things I just had to do. Bird Island was a calling,' she says now. 'My mother's people were 1820 settlers and the lights from Bird Island used to flash into one bedroom window of their house on Addo Heights. When I was small it scared me, and my mother would say, "Don't be frightened, it's just the light guiding the sailors at sea." I thought what a wonderful place this Bird Island must be. So, when I met Mr Patterson I had to grab my chance.

'I have never tried to vamp anybody as I did Mr Patterson. At that time, women weren't allowed to go to Bird Island. But I pestered the life out of him. When I was off duty from the museum, I used to go and visit him; I used to take him fruit and sweets. Finally he said, "If you get one other woman to go with you, I'll give you a permit to go to Bird Island." So I came home, and asked my mother to go with me. She said, "Yes, but what's Daddy going to say?" I said, "You just tell him we want to go and I've got a permit."'

'Margie has got round her Mother to go with her to Bird Island,' her father wrote. 'I am really furious with them. Willie is as bad as Margie. There is no holding them back.' But he granted them his permission. 'It was November 1936, and we were in Port Elizabeth, packed up and ready to go,' Margie recounts. 'The evening before we were due to leave a wire came from my father saying he was coming to Bird Island with us. But he had no permit. I went to the port captain in PE, howling blue murder, because I thought that after all this time he was going to stop me going. I wanted to get in the tug and go before he arrived, but the port captain said, wait a bit, he would see what he could do. Daddy's train arrived at about 8 a.m., it was pouring with rain, and we had been at the wharf since 5 a.m. with all our own groceries, all our stuff, loaded on the tug. I didn't even want to greet my father. He was very pleased with himself, but I wasn't happy to see him at all, because I still didn't know if they would let him go. However, he got the permit and we all went together.

'We spent about three months there – it was wonderful. Daddy was bored stiff, because it was just when the Prince of

Wales was abdicating and there were no newspapers. Of course, Daddy wanted to know all about it. Oh gosh, what a hard time he gave us. I didn't care, I was on Bird Island and enjoying myself.'

'I wonder sometimes if she belongs to me with her strange ideas of pleasure,' her father wrote in his diary while he was on the island. 'What girl of her age would isolate herself to spend her time at a Godforsaken place like this, yet she is in her glory. She has learnt to shoot and is a crack shot.'

Marjorie spent all day on the mile-square island, home to 27,000 birds. She collected, skinned and observed the breeding habits of gannets, terns, albatrosses with a ten-foot wing span, and penguins. She collected marine fish, shells, plants, anything that was of interest to the museum. It was cold and she carried a pet rabbit around in her jacket.

She continued to visit Bird Island for years. It has remained a very special place for her: 'I look at my pictures,' she says, bringing out a large album of black-and-white photographs, each carefully annotated in faded ink, 'and I feel homesick.' The pictures show a smiling young girl in a simple cotton frock, amidst thousands of birds. Bird Island was Marjorie Courtenay-Latimer's private heaven, and when it was time to leave, she had collected fifteen packing cases of specimens.

'That was when I met Captain Goosen,' recounts Marjorie. 'He was captain of the *Nerine*, and he used to stop off at Bird Island to catch Belgian rabbits for his crew, when they got sick of eating fish. Captain Goosen became interested in my work and very kindly said he would bring back my packing cases to East London, one at a time. When he had finished bringing them in, he said he would like to continue collecting for the museum, so I designed this big tank in which he could keep specimens for the museum and the aquarium. Captain Goosen was charming, a fine man. I was very fond of him. He collected everything – starfish, sharks, all kinds of things. Then he would phone and I would go down to the dock by taxi to bring them back and mount them.'

It was Captain Goosen's latest find that was causing the lady curator so much stress. Eleven days had passed since Marjorie Courtenay-Latimer had sent her letter and sketch to J. L. B. Smith, yet she had still received no response.

The letter, in fact, had been forwarded from Rhodes University in Grahamstown to Knysna, 350 miles along the coast from East London, where J. L. B. Smith and his wife, Margaret, were spending the holidays. Christmas, and then New Year, had delayed its passage. JLB was a slight man, with a weak constitution. When a friend brought round the post on 3 January 1939, he was still weak from a recent illness. Among the letters, he recognised Marjorie's handwriting. He opened the letter and read her description of the fish. He turned to the next page and saw the sketch.

'I stared and stared, at first in puzzlement,' he wrote in *Old Fourlegs*, his account of the coelacanth story. 'I did not know any fish of our own, or indeed of any seas like that; it looked more like a lizard. And then a bomb seemed to burst in my brain, and beyond that sketch and the paper of the letter, I was looking at a series of fishy creatures that flashed up as on a screen, fishes no longer here, fishes that had lived in dim past ages gone, and of which only fragmentary remains in rocks are known.' He told himself not to be a fool, but the more he looked at the sketch, at the tail, the limb-like fins and large scales, the more convinced he became of its close

Marjorie Courtenay-Latimer's rough sketch of her unusual 'limbed' fish.

resemblance to a fossil he had seen of a fish thought to have been extinct for seventy million years. 'What I suspected was so utterly preposterous that my common sense kept up a steady fire of scorn for my idiocy in even thinking of it,' he wrote.

It was a remarkable feat of mental agility. Smith had apparently taken a rough sketch by someone who was not a skilled artist, of a five-foot fish, found in the Indian Ocean off south-

ern Africa, and connected it with a fossil, a little over twelve
inches long and 200 million years old, which had been discovered in fresh water in Greenland, and which he had read
about in a scientific journal.

Margaret Smith was astounded by her husband's strange
behaviour. He was standing up, staring at the letter in complete silence. Finally, he turned to her and said, 'I know you'll
think I'm crazy – but they have found a fish at East London
of a type generally thought to have been extinct for many millions of years.'

'I *did* think he had a touch of the sun, but nine months of
marriage to this brilliant older man had taught me much, and
instead of blurting out my incredulity, I asked quietly, "What
makes you make a statement like that?" "You see that tail," he
said. "No living fish has a tail like that."' She took the letter
from him, and pointed out the date – it had been written
nearly two weeks earlier. JLB immediately started to worry.
He knew of the East London Museum's simple facilities, and
feared the worst. He sent Marjorie a telegram:

MOST IMPORTANT PRESERVE SKELETON AND GILLS
FISH DESCRIBED

Smith knew he had to see the creature in order to confirm his
suspicions, but for some inexplicable reason he didn't immediately set out for East London. In *Old Fourlegs*, he claims he
was tied up with marking University of South Africa exam
papers and could not responsibly leave the task. He also
feared that it might not be the incredible and extraordinary
discovery he hoped for. The same day as he received
Marjorie's letter, he wrote a longer reply, again urging her to
save the fish's soft parts, and confiding his suspicions as to its

provenance: 'From your drawing and description the fish resembles forms which have been extinct for many a long year, but I am very anxious to see it before committing myself ... Meanwhile guard it carefully, and don't risk sending it away.' He spent the rest of the day and night, until the post office opened the next day and he could try to telephone her, in a feverish state, with the sketch churning round and round in his brain.

In East London, Marjorie had almost given up hope of hearing from Smith. When eventually she got the wire, urging her to save the viscera, it was too late. Thirteen days had passed since the fish had first come to her, and the taxidermist, Mr Center, had long ago destroyed its internal organs and tissues. By mounting it, however, he had managed to save the internal skeleton and the skin which, by this time, had been turned brown by the formalin. When Smith managed to reach Marjorie on the telephone the next day, he stressed the importance of the innards, and sent her off to the municipal dump to try to recover them. Poor Marjorie had no luck: the rubbish of East London was dumped in the ocean.

Smith spent the succeeding weeks in a state of turmoil. The more he read, the more he became convinced that the creature was a Coelacanth (he always thought of it with a capital C), an ancient fish, whose origins date back 400 million years. His confidence was strengthened by a long-held conviction that he was destined to discover 'some quite outrageous creature ... This was so firmly fixed in my mind that just as my peculiar set of circumstances and qualifications had set the stage ready for the appearance of the Coelacanth, so in one sense had this premonition prepared me to deal with such a fantastic possibility as had now arisen, and, indeed, even while my common sense rejected it, to seek for it in an obviously impressionistic sketch by someone not an ichthyologist.'

J. L. B. Smith was experienced enough to know the importance of the discovery: it would be the greatest zoological find of the century. If, however, he announced it as such and was found to be wrong, he would be the laughing stock of the world's scientific community. Until he could get to East London, he would not know for sure. 'Those were awful days, and the nights were even worse,' he recorded. 'I was tortured by doubts and fears. What was the use of that infernal premonition of mine if it was just going to make a scientific fool of myself. Fifty million years!* It was preposterous that Coelacanths had been alive all that time, unknown to modern man.'

He wrote to K. H. Barnard of the South African Museum, cautiously expressing his beliefs. The reply was prompt and incredulous. Smith looked again and again at Marjorie's letter and sketch, as if to reassure himself. He wrote to her again:

> *Knysna*
> *9th January 1939*

Dear Miss Latimer,
Your fish is occasioning me much worry and sleepless nights. It is most aggravating being so far away. I cannot help but mourn that the soft parts of the fish were not preserved even had they been almost putrid. I am sorry to say that I think their loss represents one of the greatest tragedies of zoology, since I am more than ever convinced on reflection that your fish is a more primitive form than has yet been discovered. It is almost certainly a Crossopterygian

---

*It is now believed that the fossil record died out as long as seventy million years ago – at the time of the last great extinction.

allied with forms that flourished in the early Mesozoic or earlier,
but which have been extinct for many millions of years.
Comparatively little is known of the internal structure of such
fishes, naturally nothing of the soft parts, since fossil remains are
all that help us to know what they were like. Your fish has the gen-
eral external features of a Coelacanthid, fishes common in early
times in northern Europe and America. Whether or not it is a new
genus or family I can determine only on examination, but I feel
sure that it will make a great sensation in the Zoological world . . .

To honour you for having got this wonderful thing I have provi-
sionally christened it (to myself at present) Latimeria chalumnae
and it may even be a new family.
Kindest regards,
Yours sincerely,
J. L. B. Smith

'I was terribly upset by that letter,' recalls Marjorie. 'I was
always terrified of doing anything wrong for JLB. I phoned
and told him that if he had come earlier, nothing would have
been thrown away, but I knew it was all my fault, and I have
had to suffer that ever since.' Letters flew back and forth
between East London and Knysna, unhindered now by
inconvenient holidays. JLB wrote again to Barnard in Cape
Town, but never asked for assistance. 'At no time did I look
on it as anything but my own,' he explained in *Old Fourlegs*.
'There was no question in my mind that I had to take the
full responsibility for the decision of the identity of the
creature ... I had deliberately chosen to carry the terrible
responsibility myself, making it indeed my own funeral.' He
wrote to Miss Latimer, authorising her to offer the trawler
people £20 for another perfect specimen. She sent him some

The proud young curator, Marjorie Courtenay-Latimer with *Latimeria chalumnae.*

scales, which reinforced his confidence in his classification.

Eventually, on 16 February, JLB, a slight man who seemed almost to radiate energy, and Margaret Smith, his taller, pregnant wife, arrived in East London. It was pouring with rain, and they went directly to the museum to see Marjorie Courtenay-Latimer and the coelacanth.* According to her diary, the curator had walked to work early, and had been at the museum since 6 a.m., in a state of great excitement and agitation. Smith was ushered into the inner room where he saw the fish for the first time, sitting on Marjorie's large

*In his book, Smith wrote that Courtenay-Latimer was out when they arrived. This she refutes, vehemently: 'That always annoyed me, because I had been so anxious for him to come, and then I think it was put into his book that I was out when Prof. Smith came – as if I was going to go out shopping! I never went shopping; I had little time to shop.'

mounting table: 'Although I had come prepared, that first sight hit me like a white-hot blast and made me feel shaky and queer, my body tingled. I stood as if stricken to stone. Yes, there was not a shadow of doubt, scale by scale, bone by bone, fin by fin, it was a true Coelacanth.'

II

EX AFRICA SEMPER
ALIQUID NOVI

Of all the fish that could have possibly, miraculously, 'come back to life', the coelacanth was by far the most interesting, and Smith was well aware of its importance. The one-time existence of coelacanths had been known for almost exactly a hundred years by the time Marjorie Courtenay-Latimer came across her beautiful blue fish on 22 December 1938. In 1839, the Swiss scientist, Louis Agassiz, had described the fossil of an unusual fish tail found in a road cutting through the Permian marl slate of Durham in the north of England. He noted that the fin rays supporting the tail were hollow, and as a result, coined the name *Coelacanthus* (from the Greek for hollow spine) *granulatus* (after the tubercular ornamentation on the surface of its scales).

In the succeeding years, a plethora of similar fossils was discovered from excavations around the world: in Germany, England, the United States, China, Brazil, Madagascar and Greenland. Their defining characteristics were their hollow spines and strange lobed fins, but apart from these, it appeared that coelacanths came in all shapes and sizes. Some were fat, others were thin; they had large bodies and small tails, or large tails and small fins. They ranged in size from a couple of centimetres to a massive three metres. The most ancient specimen, known as *Diplocercides*, was found in Devonian rock, making it between 375 and 410 million years old. The most recent, of the genus *Macropoma*, only a foot long, was found in freshwater Cretaceous deposits – dating back some seventy million years – in Europe and Asia. No later coelacanth fossils have come to light, and until 1938, it was assumed that they, like dinosaurs and the majority of species, had perished in the great extinction at the end of the Cretaceous period.

Using the delicate, intricate fossil records, palaeontologists were able to make detailed reconstructions of what they believed the fishes were like, at least in terms of external and skeletal appearance. And the more they found out about the coelacanth fossils, the more excited they became.

This was in part due to a fundamental revolution in biological theory. In 1859, *On the Origin of Species* was published, and in it and his later *Descent of Man*, Charles Darwin told a wondering world that man was created by a process of evolution by natural selection; that our ancestors were monkeys, their ancestors were reptiles, whose ancestors, in turn, were fish. While gaps remained in the family tree, however, there was still a basis for scepticism. The Church maintained its anti-evolutionary position: the onus to prove the theory of evolution lay with science. Suddenly every creature, past and present, acquired a new fascination, as scientists strove to work out where each of them fitted in the evolutionary chain, and so to support Darwin's theory. Bit by bit, they began to piece together the evidence, using fossil remains of extinct

Fossil coelacanth: *Coccoderma suevicum*, 150 million years old, thirty-two centimetres long.

animals to support their case. Foremost among these 'missing links' was the mechanism by which fish walked out of the sea and began to colonise the land, literally the first step taken in the history of evolution. If evidence could be found for a tie between sea-dwellers and land-dwellers, it would help to shore up the case for evolution, and undermine the creationists. By the 1930s, when the coelacanth's continued existence came to light, evolution was still a hot topic.*

Prior to the Devonian period (between 410 and 360 million years ago), nothing lived on land, save for a few spiky, low plants, some scorpions and other insects. The earth was

---

*Even as recently as ten years ago, a survey indicated that only half of American adults believe that the theory of evolution has any basis in fact.

congregated into large continents completely different from the ones we know today, and in a constant – albeit slow – state of change. There were massive freshwater lakes, and while the land was a bare, desolate place, these lakes and oceans writhed with life. There were creatures of all forms and sizes, most of which would be unrecognisable to us today: small, flat, heavily-armoured, jawless ostracoderms (bone-skinned fish), which dwelled on the ocean floor, their mouths constantly open in search of food; giant nautilus, the size of a man; sea scorpions larger than lobsters; the first jawed fish, the placoderms, which were sometimes several metres long; and primitive giant sharks, even then fearsome predators.

It was during this period, often called the Age of Fishes, that the first bony fishes, the vertebrates, appeared on the scene. Even these would look unfamiliar today, with their encasements of heavy armour, protection from the omnipresent predators. The vertebrates were divided into two groups: the ray-finned fishes, or *Actinopterygii*, with the single dorsal fin and paired pectoral and pelvic fins common to most modern fishes; and the lobe-finned fishes – the coelacanth, the lungfish, and the rhipidistian – whose fins appeared to sprout from the end of fleshy, limb-like lobes, almost like toeless legs. These were known as *Sarcopterygii* (from the Greek *sarco* meaning fleshy, and *pterygii*, wing or fin), and were characterised also by their extra dorsal fin. Many were predators and at times, it is clear, they were extremely abundant: shale from a Triassic swamp uncovered during excavations for a new library at Princeton University contained an average of a dozen coelacanth fossils per square foot.

Some time towards the end of the Devonian period, a single species of freshwater lobe-finned fish evolved legs. In its

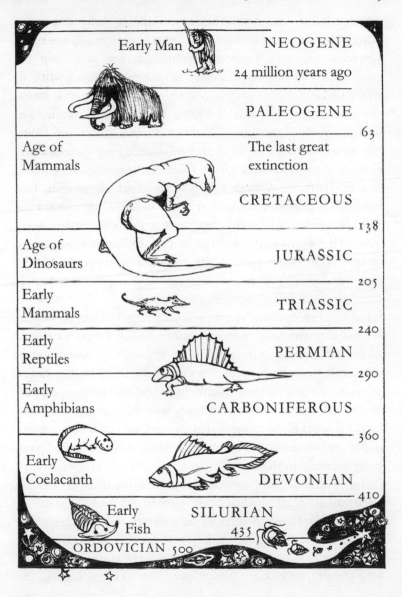

Early Man

NEOGENE

24 million years ago

PALEOGENE

63

Age of
Mammals

The last great
extinction

CRETACEOUS

138

Age of
Dinosaurs

JURASSIC

205

Early
Mammals

TRIASSIC

240

Early
Reptiles

PERMIAN

290

Early
Amphibians

CARBONIFEROUS

360

Early
Coelacanth

DEVONIAN

410

Early
Fish

SILURIAN

435

ORDOVICIAN 500

new guise of *Ichthyostega* (literally, walking fish) it crawled out of the water to conquer the land – this much scientists agreed upon. What was not so certain was which of the group evolved into *Ichthyostega*: the lungfish, rhipidistian, or coelacanth?

Palaeontologists subjected their fossil samples to minute scrutiny under ever stronger microscopes. First one, then the next, then the third was pronounced to be our fishy great-grandfather. They pored over coelacanth, lungfish and Rhipidistia fossils for signs of gills, or heart, or anything that could have preceded those first gulps of air – but without the soft parts in front of them, they were never going to know for sure. The appearance on the scene of a living *Latimeria* opened a fresh chapter in the search for our oldest living ancestor.

J. L. B. Smith circled the coelacanth several times, in silence. It was a portentous moment: after nearly seven weeks of agony and turmoil, during which he had thought and dreamed of nothing but the fish, he was eventually in its presence. His premonition had been confirmed: he was indeed the first person to recognise an extraordinary and important new creature. He went close to the fish and stroked it. Then he turned to Marjorie and said, 'Lass, this discovery will be on the lips of every scientist in the world.'

'I was amazed, and surprised and delighted, above all, that at last it had been identified and named. I asked Dr Smith all kinds of questions, which he answered only vaguely,' she recalls. 'He sat down and said, "To think that a fish of this age could exist!" I asked, "How old is it?" He counted the rings on its scales, and said it was about thirty-three years old, but its origins were more like seventy million years old. My mind boggled. So I had been on the right track. It was a living fossil.'

The night after Smith had identified the coelacanth was a very restless one, according to his wife: 'No sooner did I get to sleep than he'd waken me and ask, "Tell me, lass, I'm not dreaming, am I?" . . . Or a little later, "Sorry to waken you – but I'm not crazy, am I?"'

J. L. B. Smith knew he had to wait for official approval from his fellow scientists. He planned to announce the discovery in the leading British journal, *Nature*, and at the same time to give the fish its scientific name. What he didn't anticipate was the immediate interest and enthusiasm it would generate in the lay public. News leaked out when a journalist and photographer from the *East London Daily Dispatch* turned up (Marjorie, always conscious of what was best for her museum, had alerted the paper that the Smiths would be visiting that day). JLB granted an interview, but initially refused to let the photographer, Mr Adams, take a picture: he was worried that someone else might jump in and name the fish before he had published his report in a scientific journal. After Marjorie pleaded with him, he relented, allowing Adams to take one picture, on the express understanding that it was to be published only in the *Daily Dispatch*. Perhaps inevitably, Mr Adams acted in true paparazzi fashion and sold the picture around the world – even the East London Museum had to pay him two guineas for a print.

The scientific piracy Smith feared, however, never materialised. The *Daily Dispatch* interview appeared on 20 February, the same day as the coelacanth was put on display at the museum. From early in the morning, the curious queued around the block to see the zoological sensation that had been found off their coast. But, while the find had fired the imagination of the public, adding a new word to their lexicon, it was still greeted with a degree of scepticism by the scientific world. Several days later the British Museum – at that time God of the natural history world – telephoned Marjorie at her small museum, to ask if she was positive that the fish hadn't been dead, and just capsuled in mud for tens of millions of years before Captain Goosen trawled it up. She replied that she was sure. What proof did she have? 'So I said that at 11.30 it was blue in colour, but by 5 p.m. it had faded to a dirty grey. They asked again if I was positive, and I said, "Yes, for the last time."'

On 22 February, the East London Museum sent the fish by train, with a police guard, to Grahamstown. It was taken to Smith's house and put in a special room. 'It had a curious, powerful, and penetrating odour, an odour that in the coming weeks was always to pervade our lives, awake or asleep,' Smith wrote. His household was drilled on fish protection measures; it was to be left alone in the house at no time, and in case of a fire, the fish should be the first thing to be saved. Smith's every waking and sleeping moment was consumed by the coelacanth.

After a preliminary examination of the fish, he sent a photograph and report describing the external features of the coelacanth to *Nature* in London. The article began with the words '*Ex Africa Semper Aliquid Novi*' – there is always something new out of Africa (Pliny). The day it came out, Smith's claim to the naming of the fish was sealed. It would for ever remain *Latimeria chalumnae* J. L. B. Smith. The species would for always share a name with a pint-sized young woman from East London, and an eccentric, obsessive Grahamstown scientist.* The *Nature* article also put an end to most of the scientific scepticism. On 16 March, J. R. Norman from the British Museum presented a paper on Smith's article to the Linnean Society in London – the same august body of scientists to whom Darwin had first revealed his theory of evolution – thus giving it a further official stamp of scientific approval.

Since Darwin coined the term 'living fossil' to describe the living 'remnants of a once preponderant order', previously known only from fossil records, there had been a costly scramble to find examples of these archaic organisms. Darwin had proposed that living fossils would be found in the ocean depths, which had remained relatively unravaged by the fundamental environmental changes that, he asserted, had been the driving force behind evolutionary change.

It was to explore and survey these depths that a major three-and-a-half-year expedition was put together by the Royal Society, the British Admiralty, the Treasury and the British Museum. In December 1872, HMS *Challenger*, a converted Navy warship, set sail from Portsmouth harbour with a crew of 240 sailors and scientists. The once mighty vessel had been stripped of all but two of her guns, and laden instead with thousands upon thousands of sample bottles, microscopes and vast vats of pickling alcohol. Among the main aims of the expedition's leader, C. Wyville Thomson, was to

*'When Dr Smith wrote to say he had named the fish after me,' recalls Marjorie Courtenay-Latimer, 'I said I thought it should have been named after Captain Goosen, who had brought it for me; without him, there would have been no coelacanth. "But you ultimately saved it for science," he said.'

bring back examples of living fossils, that Darwin had predicted so confidently would live in the depths they were set to
explore.

The ship circumnavigated the globe, trawling the oceans
and seas from the tropics to the poles, with giant nets and
heavier metal dredges, which scooped up soft sediment from
the ocean floor. It was by no means a pleasure jaunt: at times
the large vessel was buffeted by giant waves, which threatened
to sweep the decks clean; four sailors and one scientist died,
two men went insane, one committed suicide, and sixty-one
deserted ship – driven to distraction by the mind-numbing
routine of years of dredging. In spite of all this, the expedition
was declared to have been a resounding success. The scientists made great advances in mapping out the topography of
the ocean floor – giving a name to a new scientific discipline,
oceanography – and they confirmed that the depths were
teeming with strange forms of life, identifying in all 4,717 new
species. They all but failed, however, in their quest for a living
fossil: they found only one, a small and not particularly interesting type of squid called *Spirula*.

In the face of the rarity of living fossils, therefore, the discovery of the coelacanth – which was not only a living fossil
but also one believed to be closely related to the lineage of
mankind – was a major event. The world's media went into a
piranha-like feeding frenzy. Articles appeared in newspapers
and magazines from New York to Sri Lanka. The long report
in the *Auckland Star* of New Zealand was entitled 'Loch Ness
Outdone'. An almost life-sized, pull-out picture was published in the *London Illustrated News*, alongside an article by Dr
E. I. White of the British Museum. Headlined 'One of the
Most Amazing Events in the Realm of Natural History in the

The East London Fish: The most startling "living fossil" ever discovered.

THE OUTDOOR WORLD

A LIVING FOSSIL CAUGHT IN THE SEA

BEST FISH STORY IN 50,000,000 YEARS

One Of Most Sensational Scientific Discoveries Of The Century

EAST LONDON'S WONDER SPECIMEN TRAWLED

Twentieth Century', it described the discovery as 'sensational', and claimed that 'this occurrence is as surprising as if one had discovered a living example of the dinosaur *Diplodocus*, the eighty-foot reptile of the Mesozoic era'. Smith, however, found sections of the article patronising, and 'not flattering to remote scientists like myself'. White had written, 'The report of this discovery was made some time ago, but was treated with justifiable scepticism by the experts – they are only too familiar with deliberate hoaxes or the misplaced enthusiasms of the uninformed to place credence in such reports until sup-porting evidence is available. (One recalls to mind the instances of the "Endfield Dinosaur", actually the remains of a luckless cart-horse, and of the "Suffolk Mammoth", the report of which sent an expert racing down into East Anglia to find nothing but the remains of fish-manure arranged in plough furrows.)'

In accordance with Darwin's proposition, White expressed his belief that the coelacanth 'almost certainly was a wanderer from deeper parts of the sea to which its kind have retreated in the face of fierce competition with the more active modern types of fishes'. This was an idea that was to become wide-spread across the scientific community, and one to which Smith reacted with derision: 'To me one glance at the Coelacanth disposed of any idea that it lived in the "inaccessi-ble depths of the ocean"; yet a number of scientists all over the world apparently accepted this with a sigh of uncritical relief . . . When I looked at that fish, even the first time, it said as plainly as if it could speak, "Look at my hard, armoured scales. They overlap so that there is a threefold thick layer of them all over my body. Look at my bony head and stout spiny fins. I am so well protected that no rock can hurt me. Of course I live in rocky areas, among reefs, below the actions of the waves and surf, and believe me, I am a tough guy and not afraid of anything in the sea ... My blue colour alone surely tells you that I cannot live in the depths. You don't find blue fishes there."'

A letter was published in *Nature* criticising JLB for naming the fish after Miss Courtenay-Latimer, since she had done science such a disservice by losing the viscera. To this, he responded in the strongest terms: 'It was the energy and determination of Miss Latimer which saved so much, and scientific workers have good cause to be grateful. The genus *Latimeria* stands as my tribute,' he wrote in *Nature*.

Smith spent every spare hour over the following months back at his house, dissecting the coelacanth and preparing his official monograph for the Royal Society of South Africa. In order to fit it in around his university duties, he would get up at three in the morning, and work on the fish until six. Then he would take a four-mile walk in the hills, write up his observations on his return, have breakfast and leave for college at 8.30. While he was there, his wife, Margaret, would type up his notes, which he would look over and correct at lunch time. She would retype in the afternoon, so that he could work again from 5.00 until 10 p.m. When his eldest son, Bob (from a previous marriage), got home from school, he would find rough sketches of fish parts with instructions attached to them, waiting to be drawn up in neat. Margaret shared with JLB each step of his dissection. 'I witnessed his agony as he found the taxidermist's iron nails or the holes they had made in the head bones. I shared his excitement when he discovered, still attached under the skin, undamaged and undisturbed, the beautiful, delicately shaped extrascapular bones that carry the sensory canal across the nape,' she recalled. It was an intense and stressful period. 'We had no social life, business and financial affairs took a back seat, and our food reached its destination over and between sheets of manuscript,' Smith wrote. 'We had no conversation, no thoughts, no ideas nor eyes, for anything except the Coelacanth, all day and all night. We could never forget it, certainly not with that smell.'

It was around that time that the coelacanth earned the sobriquet 'missing link', mainly from non-technical journalists, entranced by the story. It had the unpleasant effect of generating a torrent of letters from religious fundamentalists across the world, who were yet to be reconciled to the theory of evolution. They castigated Smith for ignoring the Bible in his 'preposterous statements' about millions of years. Did he not know that Adam was created in 4026 BC? The theory of evolution, they stormed, was evil: an anti-religious invention of the Devil put into some men's minds to enable them to divert others from the path of true thought.

Many of these letters, which Smith collected into the 'crackpot file',* came from South Africa. At that time, and even more in the succeeding decades of National Party rule, South Africa was run by Calvinist Afrikaners, who banned the teaching of evolution. Even by 1994 it was still outlawed in government schools. Smith's experience was a repetition of the reaction which had greeted another famous South African scientist, Raymond Dart, thirteen years earlier, when he published *The Taung Child*, about his discovery of relics of the first

---

*Margaret Smith apparently deemed this correspondence 'irrelevant' and from her deathbed ordered the crackpot file to be destroyed.

apeman, a true missing link to our past. He too was sent hate mail, warning that he would 'roast in the quenchless fires of hell'. Letters were written to the papers denouncing the 'satanic mischief'.

'It is a wonderful piece of irony that so many important discoveries of an evolutionary nature have occurred in South Africa,' says Professor Philip Tobias, a former pupil of Dart, who took over his mantle as South Africa's premier palaeoanthropologist. 'It was also an example of the government's schizoid mind-set. While there was a nationalistic pride that it was within South Africa that the first apeman was discovered, it was only very recently that evolution was permitted to be included – and then only as a voluntary supplement – on the school syllabus.'

J. L. B. Smith's studies progressed well. He was thrilled each time he discovered structures that matched fossils millions of years old. The board of the East London Museum, however, were becoming increasingly frustrated by their prize exhibit's continued absence. They sent a telegram to Smith, requesting its immediate return. The people of East London had been clamouring to see the fish again, and interested observers had travelled from far afield in the hope of being able to view it. Although Smith felt he had not finished his examination, he agreed to give it back. On 3 May, the coelacanth – again with its police escort – returned to its museum home. It was inundated with visitors for the next few weeks. Smith admitted to feeling a 'sense of relief' at its departure, and Marjorie Courtenay-Latimer was delighted to be reunited with her fish.

Smith's detailed monograph, comprising 106 pages of text and 44 photographic plates, was dispatched to the *Transactions*

VIC CLAPHAM

Dr. Smith said the specimen possessed features showing that it was tending to develop, and adapt itself to modern conditions.

A unique feature, discovered when Mr. R. Center, a taxidermist, skinned the fish, was that there was no skeleton. The usual bony structure was replaced by cartilage.

The mouth is proportionately large, and has even plates of sharp, catlike teeth.

The peculiar formation of the tail is also apparently an evolutionary process in forming the tail of modern fish.

SCIENTISTS' VERDICT
On reaching the port the captain of the trawler, Captain H. P. Goosen, asked Miss M. Courtnay Latimer, curator of the East London Museum, to see the fish. When she saw the great fish aboard the trawler she realised at once that it was one of a very primitive species.

NEVILLEOPTERYGII – FISH WITHOUT BACKBONE

of the Royal Society of South Africa at the end of June. It was remarkable not only for its detail, but also for the absence of a bibliography. Throughout the entire treatise, Smith did not once mention anyone else's work, something almost unheard of in scientific papers. He and Margaret had five days to recover from six frenzied months of coelacanth activity before William, their first son together, was born. 'William arrived two weeks early,' Margaret said, 'and has been in a hurry ever since.' 'My Gran,' recalls William, 'was convinced that I would have no clothes when I was born, while there were those who feared I might be born with scales. Fortunately it turned out fine. Granny saw to the clothes and there were no scales.'

Some months later, having fought off an attempt by the

museum's board to send her coelacanth to the British Museum, Marjorie Courtenay-Latimer accompanied the fish to Cape Town in a special railway van donated by South African Railways. 'That was a great experience,' she recalls, 'because every time they changed guards, they'd come and say, "The specimen's quite happy. Resting quietly." When we got to Cape Town, all the flags were flying. I thought it was to greet the coelacanth, so I waved as we drove through the streets. I later found out that there was some important foreign visitor in town.'

At the South African Museum, Mr Drury, their best taxidermist, set about remounting the coelacanth. As no one had seen a live coelacanth, and so had no idea of how it looked when swimming, he mounted it – as Robert Center had done – with the pelvic and pectoral fins pointing downwards, in a distinctly leg-like fashion.

Marjorie Courtenay-Latimer returned to East London on 3 September 1939, the day England declared war on Germany. 'Amid the chaos and the fright, all I could think of was that the coelacanth was safely down in Cape Town.'

III

INTER PISCES

J. L. B. Smith tried not to dwell on the loss of *Latimeria*'s soft parts, but their absence nagged away at him day after day, until he was forced to acknowledge his growing obsession with finding another coelacanth. 'There must be others somewhere,' he wrote, 'and at the back of my mind "a cloud no bigger than a man's hand" had formed, the forerunner of the project that came to overshadow all else in my life – the hunt for the home of the Coelacanth.'

He started to raise finance for the hire of a suitable coelacanth-hunting vessel to take him to the coral reefs and palm islands of East Africa where, he was convinced, he would discover the fish's home. He was sure that *Latimeria* was not native to the intensively fished southern African coast – if it had been, it would surely have appeared before – but was a stray that had drifted down the East African coast, carried along by the warm, south-flowing Mozambique current.

The outbreak of the Second World War, however, put his plans on hold: the world was in turmoil and until it settled down, he knew he would never be able to find a ship to take him on research trips to the Indian Ocean. He returned to his chemistry teaching, but the illustrated monograph of the coelacanth remained on his desk, a daily reminder of the great task ahead of him. It was not in his nature to give up the search; where other men might have been content with the fame and respect the first coelacanth had brought them, for Smith it was only the beginning. He *would* find another coelacanth, but for now he had to wait.

James Leonard Brierley Smith was not an ordinary man. He was a brilliant obsessive, devoted to his work, and incapable of yielding either to his own weaknesses – mainly physical – or to the defects of others. His work was his life and he involved those around him in it to the point where it became their life too.

He was born on 26 September 1897, in Graaff-Reinet, a pretty town hundreds of miles from the nearest coastline, in the middle of South Africa's Karoo desert, where his father was the postmaster. Of English, seafaring stock, it was Joseph Smith who introduced the young JLB to the joys of fishing: 'I vividly remember catching my first "Dassie", a bream-like fish, at Knysna,' he recalled. 'This wonderful shining thing I had pulled up from the unknown world below the water had a terrific effect on me, probably more than anything ever since. From then on angling has been a passion, a madness, sometimes even a reproach.'

JLB's mother, Emily Ann Beck, was a beautiful but cruel woman. She was convinced she had married beneath her, and took out her bitterness and frustration on her husband and family. According to the *Dictionary of South African Biography*, 'the parents had little in common with their elder son as they

J. L. B. Smith on a camping trip to Fish River in the late 1920s.

were quite unable to understand his sensitivity, his enquiring mind, and his craving for knowledge, education and culture'. JLB left home as early as he could, and soon broke off all relations with his mother and his sister, Gladys (to the degree that William Smith found out only recently that he had an aunt). He never talked of his family.

He excelled at the local schools, and in 1912, won a scholarship to the leafy calm of the Diocesan College (known as 'Bishops'), Rondebosch, Cape Town, one of South Africa's great public schools, where he was acknowledged to be a brilliant student. One day, he was shooting down a hill on his bicycle, when suddenly – too late – he saw a gate across the road which had not been there the previous day. He slammed into the gate and crushed his kidneys. He bled for a year. His lifelong battle with his health had begun.

The First World War broke out during JLB's matriculation year. He was keen to join up immediately, but as he was still too young, he went instead to read Chemistry at Victoria College, Stellenbosch – and came top out of the whole country in his end of year's exams, gathering a clutch of bursaries and scholarships. He was seventeen, but looked even younger, which indeed he did his whole life, something he regarded as an affliction rather than an asset.

He had intended to go to England to join the Royal Flying Corps as soon as he came of age at the end of 1915, but switched tack when the South African Prime Minister, General Jan Smuts, appealed for his people to enlist with the allies in German East Africa, now Tanzania. 'So instead of learning to soar through the skies, I became an earth-bound, foot-slogging infantryman,' he wrote. It was an horrendous period: the men lived and fought an ill-judged campaign under dreadful conditions. Smith was always a slight man, and despite his considerable sporting prowess (he was the best golfer at Stellenbosch, despite playing with only one club, and he later played rugby for his college), he was not strong. In German East Africa, he succumbed to all sorts of grim and debilitating diseases: malaria, dysentery and acute rheumatic fever. He nearly died in a Kenyan hospital and ended up being shipped back to South Africa with a permanent disability discharge. For the rest of his life, his constant battle with his wartime illnesses would preoccupy him.

JLB returned to Victoria College in 1916 to complete his BA. After another two years, during which time – never one to belittle his afflictions – he writes of being wracked by fever and more often ill than not, he gained his M.Sc. with distinction in Chemistry. He worked hard and inevitably came top in exams, but he was not a dull boy: he loved Shaw and Shakespeare, and was blessed, according to his close friend, E. G. Malherbe, with 'a fertile imagination'.

Malherbe was a fellow member of the 'Heavenly Quartet', a group of four smart men, all of whom would go on to great things (Malherbe to the directorship of the National Bureau for Educational and Social Research). In Malherbe's memoirs, *Never a Dull Moment*, he describes the 'most outrageous' student pranks that they devised and carried out. One involved putting all the university clocks forward half an hour: when none of their teachers turned up on time, the students were automatically granted a half-day holiday. According to William Smith, it was the 'precise planning' aspect of the pranks that appealed especially to JLB.

He was awarded a scholarship to study abroad, and in 1919, he went up to Selwyn College, Cambridge, for his Ph.D.

There he carried out research on mustard gases and photo-sensitising dyestuffs; he also travelled widely and learnt fluent German. JLB returned to South Africa four years later, and was appointed Senior Lecturer – and later Associate Professor of Organic Chemistry – at Rhodes University, Grahamstown, where 'Doc', as he was known, soon gained a reputation as a brilliant – if irascible – teacher, completely dedicated to his work. His students still remember his quick movements, the measured, almost pedantic way he delivered his lectures, and his habit of not looking directly at whoever was talking, then swinging round suddenly to hold the speaker in a penetrating stare.

He never, however, let go of his fondness for fishing. It led naturally to an interest in fish, and for him, there was little between an interest and an obsession. He began to spend every available minute trying to identify the fish he found. In those days there was little in the way of available textbooks, so, characteristically, he devised his own numerical system for identifying and classifying fishes. 'It took all my spare time for more than a year, and its compilation involved the writing of more than a million figures, but it worked,' he explained. He made contact with the Albany Museum in Grahamstown, and started to publish short (at first) articles on fish in the museum annals. Before long, J. L. B. Smith was a name to be respected in ichthyological circles, and he was sought out by people in need of help in identification.

It was as if he had two full-time jobs – chemistry in term-time and ichthyology in the holidays – and to both he devoted boundless energy. He started to explore further along the south coast, and became the honorary curator of fishes for the smaller, provincial museums – including, of course, the

East London Museum, with its enthusiastic lady curator – visiting them regularly to classify and investigate strange fishes. He made contact with trawler companies, and spent weeks at a time at sea with them, 'often so seasick as barely able to crawl along the slippery heaving decks to scratch among the slimy rubbish shoved aside'. He was happiest out in the field, dressed in baggy khaki shorts and sandals, his piercing blue eyes screwed up against the sun, his hair short, in a boyish, almost military cut.

At the same time, Smith started to study fossil fishes. He found this 'perhaps the most absorbing of all scientific fields; but my life was already so desperately full that I dared not indulge that desire very far. Nevertheless, those weird creatures of bygone days were constantly flitting in and out of my consciousness, constantly filling me with almost an agony that they had gone for ever and could never be seen again.' His first forty years were, though he didn't know it at the time, a perfect preparation for the discovery of the coelacanth.

In 1934, a new student entered JLB's first-year chemistry class. Mary Margaret Macdonald was a composed and popular girl, and a good student, especially of chemistry. She was the youngest daughter of a New Zealand-born doctor, William Chisholm Macdonald, and Helen Evelyn Zondagh, the first woman mayor in the Cape Colony and a descendant of the Voortrekker leader, Johannes Jacob Uys. Her great-great-grandmother, at the age of fourteen, had split open a man's skull with an axe at the Battle of Blood River when he tried to crawl under her wagon. Mary was head girl and head scholar of Indwe High School in the Cape Province, chairman of the Debating Society, and captain of the netball and tennis teams. She was a talented and accomplished singer and musician, and winner of numerous prizes at local eisteddfods. She was determined to pursue a successful career, and her first step was a physics and chemistry degree from Rhodes. There she came into contact with the charismatic 'Doc' Smith.

Initially she was terrified of him, she later admitted, and it was only in her third year that she realised that he was, after all, human. As she told the writer and photographer, Peter Barnett, 'The Professor demands a standard of work and behaviour far above the capabilities of the normal man; many of his former students are today successful men. They often come back to visit him, now as friends, and laugh over all they went through at his hands in their student days. His students fell into two distinct groups, those that liked him and those that didn't. Those who liked him were, in his own opinion, those who Worked. Capitals are used for the word "Work". He made things difficult for the others and would waste no time on them.'

In her second year, JLB announced that he was going to marry her. 'Oh no, you're not,' she replied. After she graduated, however, he trailed her to Johannesburg and insisted she marry him. ('And Dad always got his way,' says their son William.) At the time of their marriage, JLB told his young wife that while he couldn't promise her happiness, he could promise that she would never be bored. Their birthdays were both on 26 September, and until he was fairly grown-up, William believed that all parents shared birthdays.

Margaret Smith (she started using her second name after her marriage: 'While Mary Macdonald sounded good, Mary Smith was grim,' she insisted) was an intelligent, compassionate woman, who made a friend of everyone she met. She was handsome, with a strong face, and deep grey eyes. At twenty-one, she was nineteen years younger than her husband, and perhaps to disguise the age gap, he insisted she wore no make-up, and pinned her dark hair in a severe bun. He also made her give up music, one of her great passions. 'Music stirs the emotions,' he used to say, 'so it has no place in my life.'

Margaret Smith became an instant stepmother and great friend to his children by a previous marriage, Robert, Cecile and Shirley, who were only a few years her junior.* She devoted her life to JLB and his work, becoming the perfect foil for her difficult husband. She called him Len (his second name was Leonard, but everyone else called him JLB or Doc), and where he was physically weak, she was strong; she matched his short temper with patience, his distance with warmth. 'It was a helluva marriage,' recounts William Smith. 'They had incredible respect for each other. Each had what the other needed, and they were happy together – as far as Dad could be happy. I am not sure that people like him can be happy; perhaps that is what makes them great.'

From the moment of their marriage, they became a team, and until his death, they worked, lived and dreamed together. It cannot always have been easy: JLB ran his life with an intensity of purpose that permitted neither distraction nor dissension. While Margaret was an extremely intelligent and capable woman, she always had to play second fiddle to her

---

*There is no mention of the first Mrs Smith, née Henriette Pienaar, in the extensive Smith archives. By all accounts, it was an ill-suited match: she was the daughter of a Dutch Reformed Church Minister from Somerset West, in the Cape Province, and found her scientist husband hard to fathom. When they divorced, Bob and Cecile – known as Pats – stayed on with their father in Grahamstown, while Shirley went to live with her mother.

demanding, egotistic husband. 'A woman,' she once said, 'can be independent or indispensable ... but not both. I chose to be indispensable.' She proved her mettle during the discovery of *Latimeria*, acting as the rock against which JLB could bash his doubts and fears. After that, through the war years, she joined his fish-collecting trips along the South African coast, becoming as much of an ichthy-enthusiast as her husband.

JLB did not allow the frustration that the war had imposed on his coelacanth plans, to hinder his other activities. Every minute of every day was dedicated to work; no one was allowed to slack. Jean Pote, JLB's secretary from 1966 until his death, describes a 'very hard taskmaster. He didn't allow tea breaks; if we wanted tea, it had to be brought to us at our desks, so we could keep working.' She recalls how he refused to hire anyone who smoked or wore perfume, and insisted that correspondence was dealt with while it was still fresh: 'Letters are like fish,' he used to say. 'If you leave them more than three days, they begin to smell.'

According to his son, William, 'He was very impatient and impossible to live with. He was single-minded to the point of being unbelievable, and while this was a formula for success, certainly, it was not the route to happiness. As a child, living with that intellect was very difficult; I could never win. We were always fighting: if we had not started fighting within half an hour of my getting home from school, Mom would take my temperature. She was the pillow, the shock absorber between us. His behaviour, with hindsight, was in danger of breaking me. But it didn't, and I wouldn't change him for anything.'

There are numerous examples of JLB's extraordinary mental powers. He had a photographic memory, and could read

J. L. B. and Margaret Smith fishing on the Pinda Reef, Mozambique, in 1951.

sixteen languages and speak eight. When he went to Mozambique for the first time, he learnt Portuguese in three and a half weeks, and then proceeded to give an hour-and-a-half-long lecture without notes. During the war, when he was not fulfilling his teaching responsibilities, or hunting for new fish, he managed to produce three chemistry textbooks, which went into numerous editions and were translated into several foreign languages. His intellectual power was also amazingly versatile. 'We would go out for a walk,' recounts William, 'and come downwind of a buck grazing. Dad would motion for us to stop, then he would concentrate on the buck. Immediately, it would start to move around, clearly scared. Then he would turn his mind off again, and it would go back to grazing. It was incredible.' Another time he recognised, from a distance of fifty yards, a man he had never met, the son of a fellow classmate he had not seen for fifty years. The shape of his skull, apparently, had been a dead give-away.

But while his mind was strong, his body was weak. At the time of his marriage to Margaret in 1938, he was given less than five years to live. But he was determined not to give in to death, and so he didn't. Instead, he developed a proactive method of fighting his illnesses. He walked for long distances every day (it is estimated that in the last twenty-five years of his life alone, JLB walked the equivalent of twice around the world). He also changed his diet. Using his knowledge of chemistry, he analysed how the stomach worked, what got digested and where, and came up with one of the first food-combining diets. He refused to mix his proteins and carbohydrates: he never ate meat with vegetables, or bread with butter or cheese. People thought he was crazy. His sandwiches, remembers Jean Pote, used to consist of two pieces of cheese with some apple wedged between them. As Peter Barnett, who accompanied the Smiths on one of their fish-collecting expeditions up the East African coast in the 1950s, explained, 'Food is very important to the Smiths, and as they train like prize fighters, there is no room for compromise as far as their diet is concerned. Basically it is an excellent diet, but it eliminates eating for pleasure and places everything on a scientific basis.'

By the end of the war, Smith realised he could not maintain his exhausting double life. He and Margaret had become 'just machines', working all hours of the day. While he loved chemistry and was a brilliant teacher, his heart was with his fish. In September 1945 he was approached by a complete stranger, Bransby Key, to write a popular book about fishes, and offered an advance payment of one thousand pounds. He immediately handed in his notice to the Department of Chemistry, and soon afterwards successfully applied to the newly established Council for Scientific and Industrial Research (CSIR) for a research professorship in ichthyology. The university moved him into one of the old military buildings, a tiny wood-and-corrugated-iron shack, which was redesignated the new Department of Ichthyology. From then on, Smith was paid to pursue his passion.

Eight years had passed since the discovery of *Latimeria*, and he was impatient to start his quest for the coelacanth's home, and to find for himself another specimen – with its inner parts intact. He started trying to get funds together for a major pan-African expedition, nicknamed ACME (African Coelacanth Marine Expedition). The committee, composed of South African scientists with a broad range of interests, held endless meetings, and discussions raged as to the proposed breadth of the expedition's aims. But by the beginning of 1948, it was clear that ACME had 'fizzled out'.

It was not, however, in J. L. B. Smith's nature to give up. He was completely dedicated to his mission, and one way or another he was going to accomplish it. He tried another tack; if he couldn't go to the coelacanth, he would get the coelacanth to come to him. He had already been contacted by a number of people who claimed to have seen a coelacanth in different places along the southern African coast. In a few cases, the descriptions they sent were likely enough to fill JLB with optimism. Surely, he believed, if he could just spread the word more widely, it would only be a matter of time before another specimen turned up. He persuaded the CSIR and Rhodes to guarantee a reward of £100 to the finders of the first two coelacanths. Marjorie Courtenay-Latimer arranged a special coelacanth exhibition at the East London Museum, and gave the money raised to JLB, who used it to print thousands of leaflets. These leaflets (see page 56), with a picture of the fish and the offer of a reward, were translated into French and Portuguese, and distributed up and down the East African coast by every possible means. Smith had high hopes for their success.

£100
REWARD

Look carefully at this fish. It may bring you good fortune. Note the peculiar double tail, and the fins. The only one ever saved for science was 5 foot (160 cm) long. Others have been seen. If you have the good fortune to catch or find one DO NOT CUT IT OR CLEAN IT IN ANY WAY *but get it whole at once to a cold storage or some responsible official who can care for it, and ask him to notify Professor J. L. B. Smith of Rhodes University, Grahamstown, Union of S. A., immediately by telegraph. For the first 2 specimens £100 (10,000 Esc) each will be paid, guaranteed by Rhodes University and by the South African Council for Scientific and Industrial Research. If you get more than 2, save them all, as every one is valuable for scientific purposes and you will be well paid.*

In the meantime, he continued his work on South African fishes. He came to realise that, in order to put them into better context, he would have to study East African fishes as

PREMIO £100 REWARD
RÉCOMPENSE

Examine este peixe com cuidado. Talvez lhe dê sorte. Repare nos dois rabos que possui e nas suas estranhas barbatanas. O único exemplar que a ciência encontrou tinha, de comprimento, 160 centímetros. Mas já houve quem visse outros. Se tiver a sorte de apanhar ou encontrar algum NÃO O CORTE NEM O LIMPE DE QUALQUER MODO — conduza-o imediatamente, inteiro, a um frigorífico ou peça a pessoa competente que dele se ocupe. Solicite, ao mesmo tempo, a essa pessoa, que avize imediatamente, por meio de telegrama o professor J. L. B. Smith, da Rhodes University, Grahamstown, União Sul-Africana.
Os dois primeiros especimes serão pagos à razão de 10.000$, cada, sendo o pagamento garantido pela Rhodes University e pelo South African Council for Scientific and Industrial Research. Se conseguir obter mais de dois, conserve-os todos, visto terem grande valor, para fins científicos, e as suas canseiras serão bem recompensadas.

COELACANTH

Look carefully at this fish. It may bring you good fortune. Note the peculiar double tail, and the fins. The only one ever saved for science was 5 ft (160 cm.) long. Others have been seen. If you have the good fortune to catch or find one DO NOT CUT OR CLEAN IT ANY WAY but get it whole at once to a cold storage or to some responsible official who can care for it, and ask him to notify Professor J. L. B. Smith of Rhodes University Grahamstown, Union of S. A., immediately by telegraph. For the first 2 specimens £100 (10.000 Esc.) each will be paid, guaranteed by Rhodes University and by the South African Council for Scientific and Industrial Research. If you get more than 2, save them all, as every one is valuable for scientific purposes and you will be well paid.

Veuillez remarquer avec attention ce poisson. Il pourra vous apporter bonne chance, peut être. Regardez les deux queux qu'il possède et ses étranges nageoires. Le seul exemplaire que la science a trouvé avait, de longueur, 160 centimètres. Cependant d'autres ont trouvés quelques exemplaires en plus.
Si jamais vous avez la chance d'en trouver un NE LE DÉCOUPEZ PAS NI NE LE NETTOYEZ D'AUCUNE FAÇON, conduisez-le immediatement, tout entier, a un frigorifique ou glacière en demandant a une personne competente de s'en occuper. Simultanement veuillez prier a cette personne de faire part telegraphiquement à Mr. le Professeus J. L. B. Smith, de la Rhodes University, Grahamstown, Union Sud-Africaine.
Le deux premiers exemplaires seront payés à la razão de £100 chaque dont le payment est garanti par la Rhodes University et par le South African Council for Scientific and Industrial Research.
Si, jamais il vous est possible d'en obtenir plus de deux, nous vous serions très gres de les conserver vu qu'ils sont d'une très grande valeur pour fins scientifiques, et, neanmoins les fatigues pour obtantion seront bien recompensées.

well, so he and Margaret, accompanied by a team of artists, embarked upon a series of expeditions up the coast, from

Mozambique to Kenya. Before long, it became clear that the self-taught Margaret was better able to illustrate the specimens to JLB's exacting requirements, than any of the official artists, so she took over as her husband's chief illustrator and remained so for the rest of their working partnership, producing beautiful, accurate watercolours of thousands of different species of fish.

These expeditions were also an opportunity to hunt for the coelacanth. Everywhere they went, they asked about the strange fish, spreading the word and the reward poster throughout the coastal fishing villages. In Bazaruto, an island off Mozambique, JLB talked to a fisherman who claimed he had caught one several years earlier, and gave a description of a fish which fitted that of the coelacanth. This was clearly a positive indication of the coelacanth's whereabouts. Sadly, it was the only one.

At about the same time, however, on the other side of the world, the first of a series of clues surfaced, which seemed to hint at a secret existence of the coelacanth, far from the area of the western Indian Ocean on which J. L. B. Smith had focused his search. One day in 1949, Dr Isaac Ginsburg of the Department of Fishes at the National Museum in Washington DC, received a small package in the mail. A woman from Tampa, Florida, had sent him an unusual fish scale, about the size of a silver dollar and different from any scales he had seen before. The woman explained that she ran a small souvenir shop, stocked with fishy artefacts, many of which she made herself from odd bits of flotsam and jetsam – pretty shells and fish scales. One day in 1949, a local fisherman had walked into her shop and sold her a gallon barrel of strange scales. Out of curiosity, she had decided to mail one of them to the museum for identification.

Ginsburg turned the strange scale over and over. He was puzzled; he had never seen scales like that either, and he knew immediately that the fish to which they belonged was not known to live in the Gulf of Mexico, or in any American waters. He wrote back to the souvenir seller, asking for further details, but she never replied and he was unable to track her down. He became convinced that the scale belonged to an ancient fish, probably a crossopterygian and very possibly a coelacanth. In 1949, only one coelacanth was known to science – *Latimeria chalumnae* – and its scales were fully accounted for.

Smith's attention, however, was not to be diverted. *Sea Fishes of Southern Africa* was published in July 1949 and sold out within three weeks. In the acknowledgements, JLB paid fulsome tribute to Margaret: 'My wife has been my full-time partner from the beginning and has been artist, adviser, buffer, critic and secretary, and is one of the most skilful collectors of fish, with many devices of her own. She has shared with me many hardships and sustained me when my courage has failed. But for her spirit, energy and unflagging enthusiasm this work could hardly have reached completion in any reasonable time.' A new edition was rapidly prepared and was issued the following year, again to a tremendous reception.

In the early 1950s, the Smiths embarked upon another series of fish-hunting expeditions, to research further books on the fish of southern Africa. On one, to Mozambique, they took along Peter Barnett, a young English photographer, who later wrote a colourful book about his adventure.

In *Sea Safari with Professor Smith*, he describes a gruelling expedition with an extraordinary, uncompromising man. 'It was never Professor Smith's habit to praise,' Barnett wrote, 'because he said, "When a motor runs smoothly there is no need for repair, for comment. But when faulty, it has to be licked back into perfect running order." ... No matter what I did, I would become entangled in the complexities of the Professor's will ... I began to understand a little more about this man, an extreme egotist, sublime in the certain knowledge of his own intellect.'

He was impressed by JLB's immaculate punctuality: 'To such a man as Smith if one arrived a minute over an appointed time, say 3.15 a.m., then the logical conclusion must be, "Why not then arrange the meeting for 3.16 a.m. or 4.16 a.m.?" JLB used to think that what separated *Homo sapiens* from the rest was time, and he was not convinced women were *Homo sapiens*, because they were not always on time. He was, and he didn't have to wear a watch to be so.' He was also frighteningly well organised: 'Professor Smith, like Lewis Carroll, numbers every letter he writes, files everything, and had listed every article, down to the smallest box of pins, in the fifty-odd pieces of baggage we took on the expedition. His list, for example, would read "Box 48, 3/4 down right side, forceps".'

As the journey progressed up the coast, Barnett saw that the Smiths 'were quite insane about fish, but I soon came to realise that their whole inner selves were focused on the coelacanth'. He told of how they always carried around sheaves of the reward posters, and frequently saw them displayed on notice-boards in ports, where the coelacanth was known as the *Dez conto peixe* (*sic*), the hundred-pound fish.

'Both of them were always talking about it to everyone who would listen, which was everyone, because as he said, £100 talks in all languages.'

Their general *modus operandi*, on arriving at a harbour, was to make straight for the fish market, then visit netting boats and fish traps, before bargaining for the fish they had selected. When they went out collecting themselves, Margaret would row the boat, while JLB issued directions. When they reached what he thought was a suitable spot, he would toss dynamite into the water, and Margaret would dive in to collect the dead fish. JLB didn't like to get wet. 'Neither he nor Mrs Smith, I believe, enjoy their expeditions in the accepted and normal way, such as one might experience when hiking,' Barnett wrote. 'Enjoyment in their busy lives, comes from the inner satisfaction of accomplishments and hardships that they endure willingly in the pursuit of scientific knowledge.'

Barnett soon realised that he was expected to push himself to the same degree. He was sitting in a wicker chair after lunch one day, when he was pounced upon by the Professor: '"Have you any work to do? . . . Relaxation?" he said in tones of stark disbelief. "But you relax when you sleep, and you'll relax for ever when you're dead. In the meantime we've got work to do."' Everything was accomplished at top speed: 'The Professor and Mrs Smith walked fast, the theory being that not only is walking a means of transport, but one must also derive the benefit of exercise from the action so as not to waste time.'

Peter Barnett spent several hard months cataloguing the expedition, and by the end, he found himself torn between awe and incredulity at the Smiths' dedication and profession-alism. One time, JLB rebuked him for not working hard enough, and said that if he didn't pull his metaphorical socks up, he should leave the expedition. Barnett persevered, but when he left them eventually, 'the Professor did not wave'.

In 1952, fourteen years after JLB first vowed to track down another coelacanth, the Smiths embarked upon another expedition up the east coast. They were both becoming frustrated by their inability to find a coelacanth. They had scoured even the smallest fishing villages, but despite seeing their poster displayed prominently, they had encountered no one who recognised the coelacanth, apart from the one possible sighting in Bazaruto. Yet JLB was certain it was there, somewhere, waiting to be found – if only he knew the right place to look.

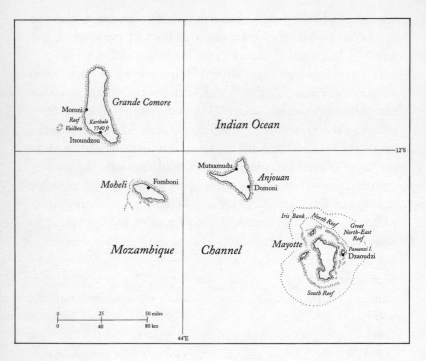

The Comoros

IV

MALANIA ANJOUANAE

'I wonder where that blasted fish is. Come on, lass, let's go to the Comoros.' The Smiths were standing at Cape Delgado, Mozambique, staring across the Indian Ocean at the point where the prevailing current divides, one section heading south along the coast to Lourenço Marques (now Maputo), Durban and eventually, to East London. JLB had often mentioned his desire to search for the coelacanth in the Comoro Islands, a remote volcanic archipelago at the head of the Mozambique Channel, half-way between Mozambique and Madagascar. The four tiny islands were Arab trading posts for hundreds of years, ruled over by a succession of battling sultans. They became a French colony in 1946, but remained a largely forgotten outpost, unaffected by the twentieth century. JLB was keen to explore their famed coral reefs, but the islands were just too far from the mainland coast to allow easy access by boat, and this time, as on previous occasions, Margaret managed to dissuade him.

The Smiths stopped off in Zanzibar for a couple of weeks of fish collecting on their 1952 expedition. At the end of their stay, they were invited by the authorities to exhibit the more interesting fishes they had found there. The show was crowded with locals, diplomats and seafaring folk, all keen to see the celebrated Professor's catches. Late in the afternoon, a friend of the Smiths' arrived, bringing with him Eric Hunt, a keen amateur ichthyologist and captain of a trading schooner. The handsome, charming Hunt, often described as a dead ringer for Errol Flynn (only shorter) was introduced to Margaret, and they started talking animatedly about fish.

As he turned to leave, Hunt picked up one of the coelacanth leaflets. 'Do you think this fish might be found in the Comoros?' he asked Margaret. 'The Comoros!' she replied. 'What makes you ask about the Comoros?' He explained that he often made the journey down the coast, trading local produce, dried fish and sharks between Zanzibar and the Comoro Islands, a few hundred miles into the Indian Ocean. 'I think there is a very good chance that it may be found at the Comoros,' Margaret told him. 'My husband certainly feels that the Comoros are about the only place left where they may occur.' Hunt gestured at the pile of leaflets. 'May I take some of these?' he asked. 'The Governor of the Comoros would certainly be delighted if one of his subjects could collect a reward of this size.'

Two months and many fishing trips later, the Smiths' ship docked at Zanzibar on its journey back to South Africa. Margaret went ashore to visit the bustling market, and as she was returning to the wharf, was hailed by Hunt on his trim schooner, the *N'duwaro* (marlin or swordfish in Swahili). He had recently returned from the Comoros, he reported, where he had shown the leaflets to the French Governor who had seemed very interested and immediately organised for them

Eric Hunt and the reward poster, Comoros, 1952. He suffered from
dry skin, and was frequently photographed licking his lips.

to be distributed throughout the archipelago by native run-
ners. Hunt was full of enthusiasm. It was just the kind of
quest that appealed to his adventurous, romantic side, and he
was keen to be involved.

Eric Ernest Hunt was an adventurer in the purest sense of
the word. Born in 1915 to a respectable family from East
Sheen, a leafy south-western suburb of London, he went to

live in East Africa in 1935. He worked as a motor mechanic for a time, ran a ferry service on Lake Victoria, and then, at the outbreak of the Second World War, enlisted in the Royal Electrical and Mechanical Engineers. He served in Abyssinia and East Africa until the end of the war, receiving citations for his bravery.

In 1946, he embarked upon a new career as a sea trader. He based himself in Zanzibar and, with progressively larger vessels, plied the western Indian Ocean coast and offshore islands with tea and coffee, spices, cloths and cloves. He was well liked wherever he went, a handsome, somewhat shy man with a quiet charm and serious demeanour. His boats were kept in immaculate condition, and he treated his crew with respect and generosity. Like Smith, his passion for fish arose from a love of angling. Over the years, he became increasingly interested in collecting fish for his private aquarium, and studying their behaviour. The chance encounter with the Smiths in 1952 and his involvement with the coelacanth provided the kind of excitement he craved.

He plied Mrs Smith with more questions about the coelacanth: how he would recognise it, and what he should do if he found one and had no formalin. To the latter, she replied that he should salt it, in the same way that he salted sharks. As he waved goodbye, he called out, 'OK, Mrs Smith – and when I get a coelacanth I'll send you a cable.' They both laughed.

The *Dunnottar Castle*, the large Union-Castle liner which had carried the Smiths down the coast, docked at Durban on Christmas Eve. It was stiflingly hot, but despite the holiday torpor, a buzz of excitement greeted the arrival of South Africa's most famous scientist. Friends and journalists swarmed aboard the ship to hear news of the expedition. JLB was handed a fistful of telegrams, one of which, he noticed, had been forwarded from Grahamstown and was marked urgent with a red sticker. He was talking to a reporter at the time, and during a lull in the conversation, he casually slit open the urgent message. His reaction was remarkably similar to that on the day he received Marjorie Courtenay-Latimer's first letter. Suddenly, he was on his feet, speechless. 'Two words stood out: "Coelacanth" and "Hunt",' he wrote in *Old Fourlegs*. Margaret Smith looked up in alarm and took the telegram.

REPEAT CABLE JUST RECEIVED HAVE FIVE FOOT SPECIMEN COELACANTH INJECTED FORMALIN HERE KILLED TWENTIETH ADVISE REPLY HUNT DZAOUDZI.*

Smith's mind started to fly. He had no idea where Dzaoudzi was, but he knew he would have to get there, and fast, if he was to prevent a repetition of the disastrous loss of *Latimeria*'s soft parts. It was Christmas Eve, four days after the fish had been killed. Time was running out.

Smith asked a young officer to find out where Dzaoudzi was. The officer ran off and was back in a minute with the information that Dzaoudzi was a small islet near the island of Mayotte in the Comoros. It was the Comoros after all! Now Smith had to find a way of getting there: no commercial airline flew to the islands and a boat would take far too long. His only option was to charter a plane.

---

*Hunt had sent an earlier cable to Smith at Grahamstown, which had been intercepted by his secretary, who relayed the information to the Professor on the *Dunnottar* later in the day.

He stationed himself at the bridge of the *Dunnottar Castle*, commandeered the only phone, and got down to work. First he drafted a reply to Hunt:

IF POSSIBLE GET TO NEAREST REFRIGERATION IN ANY CASE INJECT AS MUCH FORMALIN POSSIBLE CABLE CONFIRMATION THAT SPECIMEN SAFE. SMITH.

He then tried to reach the President of the CSIR, Dr P. J. du Toit – but found he had already left his office for the holidays and could not be traced. Smith mentally ran through the roll of cabinet members, some of whom he knew, and decided to try Eric Louw, Minister of Economic Affairs: 'From the lines on his face his duodenum probably twisted the same way as mine, and I had a suspicion he might look on Christmas time with the same jaundiced eye as myself,' Smith wrote. However, Louw was in the USA. Smith tried Donges, an old Stellenbosch contemporary and Minister of Internal Affairs. The post office eventually managed to contact him as he got off a train in Cape Town, but he said that much as he would have liked to help, regretfully he didn't think he would be able to do much from Cape Town on Christmas Eve. He suggested JLB contact the Prime Minister, D. F. Malan.

Smith wasn't keen to take that route. Several years earlier he had tried to talk to the then Premier, General Jan Smuts, to try to secure his help in getting to Walvis Bay, on the southwest African coast. Millions of dead fish had just been washed up on the beach, the victim of a red tide of planktonic organisms, and JLB was desperate to collect specimens. He went in person to Smuts' official Cape Town residence to request an urgent meeting, but Smuts refused even to see him. From that time onwards JLB had nursed a healthy distrust of Prime Ministers.

So, instead of calling Malan, Smith tried to contact the Ministers of Transport and Defence and the head of the armed services. He was frustrated at every attempt: they either could not be reached, or could not help. By this time it was Christmas Day, and virtually impossible to get hold of anyone. ('Why on earth did Coelacanths want to turn up just before Christmas?' he wrote.) As in 1939, Smith was unable to sleep; he was tortured by twin fears – that the coelacanth might be decomposing as the days drifted on or, worse still, that it might not in the end turn out to be a coelacanth. He was staking his career, he realised, on the word of a layman who had never seen a coelacanth before.

On Boxing Day another cable arrived from Hunt:

CHARTER PLANE IMMEDIATELY AUTHORITIES TRYING TO CLAIM SPECIMEN BUT WILLING TO LET YOU HAVE IT IF IN PERSON STOP PAID FISHERMAN REWARD TO STRENGTHEN POSITION STOP INSPECTED* FIVE KILO FORMALIN NO REFRIGERATOR STOP SPECIMEN DIFFERENT YOURS NO FRONT DORSAL OR TAIL REMNANT BUT DEFINITE IDENTIFICATION HUNT

JLB's mind was whirring: 'I realise now that I was in a state of mind that is termed "Possessed",' he wrote. He feared for his fish – and although he was convinced of his claim to ownership, he knew he would be able to do little if the French decided it was rightfully theirs. A plane was needed – and

* Presumably a misprint for injected.

immediately. Smith was desperate to see the fish for himself.

In his desperation, he knew that there was only one chance left; he had exhausted all other possibilities and would now have to overcome his aversion to Prime Ministers and beg the help of Dr Daniel François Malan, the fundamentalist, anti-British, deeply religious Prime Minister. After JLB's experience with Smuts, a professed lover of science who would not even see his country's most eminent scientist, let alone help with a domestic flight, he did not have great hopes for Malan, who would have to give so much more – an international plane trip with all the political complications that might ensue. Still, he realised he had no option but to try. Smith engaged the services of the local MP, Dr Vernon Shearer, and together they telephoned the Prime Minister's holiday cottage near Cape Town.

Shearer got through to Mrs Malan, who told him that her husband was already in bed and that she would not disturb him. '10.30 p.m. of the 26th December in the year of our Lord 1952. It was probably the lowest ebb of my life,' Smith wrote, with typical understatement. 'The sands of time were running out, fate was screwing me down to the dregs, wringing out the last drops of my spirit from the rags of my being . . . What on earth was I to do, for now there seemed no more hope?'

Suddenly the phone rang. Shearer picked it up, said a few words and called for JLB: 'Quick, Professor, Dr Malan! Dr Malan wants to speak to you!' JLB took the receiver: 'Mrs Malan here, Professor; the Doctor wants to speak to you.' Then the familiar voice came on the phone, speaking in English. 'Good evening, Professor, I have heard something of your story, but will you please give me as full a summary as possible.' JLB began, insisting on speaking in Afrikaans, even though he stumbled over some of the more technical terms. 'I gave him a brief account of the fantastic discovery of the East London fish, the tragedy of the soft parts, my long search, the recent discovery, the heat, the isolation, my fears and my needs.' Smith explained that there was a chance that it might not be a coelacanth, but that there was no question in his mind that it was worth the risk. He said that in his view it was a matter of national prestige – that South Africa had a right and a responsibility to the fish.

He talked for twelve minutes. Malan listened intently, then, when he finished, congratulated him on his Afrikaans. JLB waited. 'Your story is remarkable,' Malan continued. 'And I can see at once that this is a matter of great importance. It is too late to try to do anything tonight, but first thing in the morning I shall try to get through to my Minister of Defence to ask him to locate a suitable aeroplane to take you where you need to go.'

'As I put down the receiver,' Smith recalled, 'I felt dazed, like a man reprieved on the very scaffold, like somebody suddenly jerked from the hollows of hell to a high hill-top in heaven.' He was astounded by Malan's positive reaction, particularly to someone who was a scientist with a British name, from such a very British institution as Rhodes.

Later, Smith found out what had happened that night. When the phone rang, Mrs Malan had judged it better not to wake her husband. However, he had heard the ringing and called out from his bed to ask who it had been. She explained briefly and Malan nodded, and said, 'This man Smith is well known. Bring me that fish book.' Some months before, Smith had sent his *Sea Fishes* book to the Prime Minister, and by some divine shake of fate, Mrs Malan had brought it to their beach house for the holidays. That night Malan leafed through the book and read the coelacanth section. Then he shut it, and tapping the book said, 'The man that wrote this book would not ask my help at a time like this unless it was desperately important. I must speak to him.'

Smith worked through the night, preparing lists of things he might need for the trip: food, clothes, a Primus stove and his collecting kit – a teak box containing trays of tools, spare parts, medical equipment and fishing tackle. He also managed to find two gallons of formalin, which he added to his travelling gear. By morning he was ready. He went to the dock to wave goodbye to Margaret, who was heading homewards on the *Dunnottar Castle*, then went back to Shearer's house to wait for news from the Prime Minister. At 3.30 that afternoon Smith received a message to say that the route had been cleared, and that an Air Force Dakota would be in Durban at dawn to collect him. He sent a telegram to Hunt:

HOLD ON STOP GOVERNMENT SENDING PLANE

That night JLB managed to sleep for three hours, but he was still awake well before first light. He went to the airfield to meet the clearly sceptical Dakota crew. Smith greeted a bemused Commandant Blaauw: 'I bet when you joined the South African Air Force you never expected to command a plane sent to fetch a dead fish.' He climbed into the unlined hull of the military aircraft with his boxes, and gallons of water, which he insisted were necessary. The plane then took off on the first stage of its journey. It was noisy in the back, and Smith was uncomfortable. He was excited at the prospect of seeing the coelacanth, but also a little nervous: the Commandant had informed him that, while the flight had been cleared as far as the Comoros, they had not been able to reach anyone on the islands to alert them of their arrival. Indeed, they had not even been able to find out whether there was a usable landing strip in the Comoros. They would just have to fly there and hope for the best.

The plane stopped to refuel in Lourenço Marques, then headed north towards Mozambique Island. Later, Smith found out that it had not been exactly plain sailing getting permission for landing rights in LM. At 2 a.m. the previous morning, an Air Force ADC in Durban had called a government official in LM on a crackling line. He requested permission from his opposite number for a military plane from South Africa to pass through the territory and refuel at the air base. 'Roger,' said the LM man. 'And what is the mission of this flight?'

Durban: 'To get a fish.'

LM: 'Have I heard you right, a F.I.S.H?'

Durban: 'Yes, a fish.'

LM: 'You mean a thing with scales?'

Durban: 'Roger.'

LM: 'Do you really think our government is going to believe that? You must think our guys are stupid – can't you think of a better story for why you want to cross our territory in a military plane?'

And so the bizarre conversation continued. The official in LM eventually agreed to request permission from the Portuguese Governor General – a friend of Smith's – and was amazed when he got an immediate green light.

The crew spent the night on hot and humid Mozambique Island, off the northern coast of Mozambique: a familiar hunting ground of the Smiths. JLB passed another sleepless night, and by the time the plane took off before dawn the next morning, he was extremely wound up and nervous. He had staked virtually his whole life on this expedition, and would be ridiculed and ruined professionally if it turned out to be a false alarm. The plane flew low across the Mozambique Channel and soon sighted the first of the four Comoro Islands. From above, they looked like emeralds, smothered in lush vegetation and surrounded by an aquamarine fringe of coral and beyond, only a few hundred yards offshore, the deep blue of the ocean. Each was mountainous and on the largest, Grande Comore, Smith and the crew caught a glimpse of Karthala, the largest active volcanic crater in the world. There was little sign of civilisation – just some small villages clustered to the sides of the mountains, or sheltered beneath tall palm trees along the coast, and in the sparkling sea, tiny, speck-like canoes with two outriggers like wide-stretched arms. The most distant of the islands was Mayotte, shaped like a sea horse, with the islet of Dzaoudzi at its side. But they were still unable to raise the Comoros by radio.

The plane had started to descend, circling the island, when Lieutenant Bergh gave the thumbs-up sign; he had seen the airstrip. JLB looked out of the window, and far below saw a tiny vessel bobbing about in the harbour. He felt a surge of excitement when he realised it had to be Hunt's ship, the N'duwaro, with his coelacanth aboard.

The plane made a bumpy landing on the rutted strip, and immediately it started to pour with rain. 'The rain stopped as abruptly as if turned off by a tap, the mist parted and figures came running across the flattened coral rag,' Smith wrote. 'The door opened and through a blast of hot air I saw Hunt's face looking up at me. For a moment I could not speak; then with a rush of pent-up emotion the words "Where's the fish?" burst from my lips like an explosion.' Hunt reassured Smith that his fish was safe, and led him to the Governor's house. Smith was clearly desperate to see if it was indeed a coelacanth, but Hunt insisted that he greet the Frenchman first. For his part, Governor Pierre Coudert, resplendent in his white tropical uniform, was keen to meet the man who had managed to persuade a Prime Minister to send a plane to collect a fish.

'I have often suffered from the necessity of paying tribute to officialdom,' Smith wrote. 'But this was probably the hardest I have ever endured. It was agony and torture, and I raged inwardly, my mind a searing flame. Blast those formalities! I had not endured all that I had been through or come so far to exchange polite words with a Governor at that critical moment. I wanted only one thing, and that was to see the fish, to know if I was a fool or a prophet.'

The South Africans were introduced and offered food and drinks that were laid out on a long table. But JLB could endure it no longer; he thanked the Governor through gritted teeth and asked most respectfully if they could return after they had inspected the fish. He almost ran to the car, and was soon down at the dock and on Hunt's schooner.

Hunt pointed to a large, coffin-shaped box by the mast, which he ordered to be opened for Smith to see. A sea of kapok was covering the fish. 'My whole life welled up in a terrible flood of fear and agony, and I could not speak or move,' Smith recalled. 'They all stood staring at me, but I could not bring myself to touch it; and after standing as stricken, motioned to them to open it.'

'God, yes! It was true! I saw first the unmistakable tubercles on the large scales, then the bones of the head, the spiny fins.

It was true! Malan would not suffer for his action, thank God for that! It was a Coelacanth all right. I knelt down on the deck so as to get a closer view, and as I caressed that fish I found tears splashing on my hands and realised that I was weeping, and was quite without shame. Fourteen of the best years of my life had gone in this search and it was true; it was really true. It had come at last.'

Smith could have stayed stroking his fish for hours, and the crew would have been more than happy to stay a bit longer on the beautiful and unspoilt islands. But Smith was also nervous that he might yet be prevented from removing his coelacanth. That fear spurred him into action: he carefully lifted the fish out of its box and posed it for photographs. He examined it rapidly, noted its length (four feet six inches or 1.385 metres), the lack of first dorsal fin, and truncated tail. While it was without a doubt a coelacanth, Smith surmised it was of a different species.

As he was doing this, Hunt recounted the events leading up to the dramatic rescue mission. By the end of November 1952, Smith's leaflets had been widely distributed throughout the islands. They were greeted with great interest, and the Comorans were amazed at the sum being offered for one fish. On the night of 20 December, a fisherman named Ahamadi Abdallah,* with his assistant, Soha, took his small *galawa*

---

*In *Old Fourlegs*, Smith wrote that the fisherman was called Ahmed Hussein Bourou, but when explorer Quentin Keynes visited the Comoros in 1954, he spoke at length to the fisherman concerned, who assured him that his name was Ahamadi Abdallah, the same name mentioned in the official French report. When Keynes took this up with Smith, he met with the retort: 'French administrators were not to be regarded as the most reliable sources of information.'

Eric Hunt, J. L. B. Smith and the crew of the Dakota on Dzaoudzi, Comoros.

canoe to fish off the south-east coast of Anjouan, near the town of Domoni. He let out his long hand line, and after a few hours he hooked a big fish, at a depth of 160 metres, which he killed by battering its head. Satisfied with his catch, he returned to his village, and left the fish outside his hut, without scaling or gutting it.

The next morning, he took it to the beach to clean. However, just as he was about to start, he was approached by a local teacher called Affane Mohamed, who was having his hair cut nearby. Mohamed (who would later become the Comoran Minister of Culture) recognised the fish as being very similar to the one on the leaflet. The instructions were clear: 'Do not cut it or clean it or scale it, but take it at once to

some responsible person.' Mohamed urged Abdallah to stop
what he was doing, and took him to where the poster was displayed. The fisherman was at first unwilling to believe that
anyone would be prepared to pay such a fortune for what he
regarded as an almost useless fish, but he was persuaded at
least to see if it were true.

It was known by the bush telegraph – or Radio Cocotier as
it is called in the Comoros – that Captain Hunt was at this
time moored at Mutsamudu, on the other side of the island. It
was also known that it was Hunt who had brought the leaflets
to the islands for Coudert to distribute. Legend has it that
Abdallah lugged his precious eighty-two-pound (thirty-seven-
kilogram) cargo by foot in the sweltering heat across twenty-
five miles of mountainous terrain, but according to Keynes's
investigations, he managed to get a lift on a Public Works
truck on its way back across Anjouan from Domoni to
Mutsamudu.

Whatever way he got there, by the time he reached Hunt
the fish had already started to putrefy. Hunt immediately
recognised it to be a coelacanth and started to work out a way
to save it for Smith. After a quick enquiry affirmed that there
was no available formalin on the island, Hunt ordered his
crew to cut and salt the coelacanth, as Margaret Smith had
suggested. He promised Ahamadi Abdallah that he would
take the fish to Dzaoudzi, the seat of the Governor and the
only town with the means of international communication,
and that he would return with the 50,000 Comoran Francs
(CFA) reward, the equivalent of £100.

Hunt set sail for Dzaoudzi, with the fish and the Anjouan
football team. When he arrived, he contacted the Governor
and explained the situation. He also managed to track down

the resident French doctor, who gave him formalin to inject into the decomposing fish. Coudert immediately cabled the French scientific base in Madagascar for instructions, but – luckily for Smith – the cable got mutilated in transmission and delayed as a result of the holiday period. Had the French scientists received the correct message, they might well have claimed the fish for themselves. Instead, when Coudert heard nothing, he gave his word that Smith could have it, if he came in person to collect it. The coelacanth was won for South Africa.

It was clear that Hunt had stuck his neck out for Smith, in persuading the authorities that the coelacanth belonged by right to South Africa. In return, Smith suggested he christen it *Malania hunti*, after the Prime Minister and Eric Hunt, but Hunt demurred. He said he would prefer it if somehow the French were honoured, since it had been, after all, fished off their territory and his livelihood depended on good relations with them. So Smith fixed instead on *Malania anjouanae*, to commemorate the island where the fish was found.

Furthermore, JLB said he was prepared to offer £100 for the capture of another specimen, and that if it were to be caught in French waters, it would be offered to the French.

The Governor, Monsieur Pierre Coudert, was delighted at the suggestion of naming the coelacanth after one of his islands, and said the Smiths would be very welcome to return to study the fish on the reefs surrounding the island. According to Smith, Coudert was extremely enthusiastic and offered the visitors every hospitality. 'Madame was concerned at my lack of appetite,' he wrote. 'Right in front of me was a schoolboy's dream, an enormous cake spread with sticky chocolate icing, the mere sight of which made my liver throb.'

Rather before they politely should, Smith and the crew thanked the Couderts and after he had sent cables to Margaret, Malan and the CSIR confirming that it was indeed a coelacanth, they left for the airstrip. They had been on the ground for less than three hours, and Smith had met neither the fisherman who had caught the coelacanth, nor the schoolteacher who had correctly identified it. As JLB climbed back into the plane for the return journey, he gave Hunt £200 – £100 for the reward to the fisherman and the rest to cover his own expenses – and a handful of newspaper cuttings about the event. At 10 a.m. on Monday 29 December the Dakota took off for the return journey to South Africa. They had just got above cloud level, when Captain Letley passed a note to Commandant Blaauw, who passed it to Smith: 'Managed to intercept a message stating that a squadron of French fighter planes left Diégo-Suarez before we took off from Dzaoudzi with orders to intercept us and to compel us to turn back.' Smith turned pale. He questioned the pilots about their ability to dodge the fighter planes. They said it wasn't likely – the Dakota was too slow and too cumbersome. 'Well,' replied Smith, 'I don't know how you chaps feel about this, but I'm not going back. I don't believe they would dare to shoot us down if we refused to turn, but I would be prepared to chance that rather than turn back.' The Captain burst out laughing, and it was some seconds before Smith realised that it had been a hoax.

He went to the back of the plane, and to keep his mind diverted from his aching ears, he started to make notes about the experience. At one point he decided they all needed some coffee, and it was pure good fortune that one of the crew surprised him as he huddled in the back of the plane, trying to

fire up his travelling Primus stove. It took the intervention of
the Captain and Commandant – who told him that the fish
would be blown to bits if he continued – to stop him. He
remained convinced, however, that it would have done no
harm. Commandant Blaauw informed Smith that the flight
was costing the government £40 per hour. JLB did some
rapid sums and calculated that, assuming everything went to
plan, by the time the fish was back in South Africa it would
have cost at least £1,000 for the flight, which, with the reward
and Hunt's expenses, would make it, surely, one of the most
expensive fish of all time, equal to about £12 per ounce in
today's prices.

The crew refuelled in Lourenço Marques and at 6.45 p.m.
took off again on the last leg of their journey. They had been
awake since 3 a.m., but were full of excitement and satisfaction at their successful mission. After an initially cool start,
they had come to like and respect the crazy Professor. Smith
handed them all a piece of paper and asked them to record, in
words fit to print, their first thoughts on being called away
from their holidays for this extraordinary assignment. The
Commandant was typically dry and understated, but Captain
Letley wrote, 'The first time that I knew we were going to
fetch a fish (DEAD) was when the Orderly Officer told me.
My reply, as you requested, can not be written down.' They all
said they had enjoyed the experience, nevertheless.

They landed at Durban, exhausted, to be greeted by a bat-
tery of flash bulbs. J. L. B. Smith was the man of the hour.
He cleared the fish through customs, and spoke to the
Commander-in-Chief, requesting permission to take the
plane on to Cape Town the next day to show the coelacanth
to the Prime Minister. It was granted. Then the South African

Broadcasting Corporation reporter said that the country was waiting for a live radio broadcast from Smith: all programmes had been dislocated that evening to make time for him. His first reaction was to refuse – he had not slept properly for almost a week, and had no time to prepare such an important announcement. However, he remembered the notes he had made in the plane and asked for twenty minutes to organise his thoughts.

The broadcast began, and Smith's confidence grew. His delivery was typically measured, but he could not disguise his emotions as he started to relive the experience: when he told of weeping at the sight of the fish, tears started to fall again. When he finished, he was completely spent. The programme was later described as one of the most emotionally charged pieces of broadcasting to have ever been aired on South African radio.

After speaking to Margaret, JLB tried to get a few hours' sleep. The coelacanth lay snug in its kapok-lined coffin beside him in his bedroom at the Snell Parade Barracks, while a special detail of Zulu guards patrolled outside.

V

STAR FISH

The dramatic rescue of *Malania* was splashed across the front pages of the world's newspapers. On 27 December, the *New York Times* ran a story under the headline: 'Prehistoric Fish Believed Caught'. Three days later, the *New York Herald and Tribune* announced that 'Air Race to Save Dead Fish Stirs Scientists Here'. The *Times of Malta* reported that 'Malan Sends Plane to Collect Reputedly Extinct Fish', while the Karachi *Dawn* exclaimed, 'Missing Link Found!' The coelacanth and the Comoros had entered the global lexicon.

On 30 December, Smith and his *Malania* were up and in the air before dawn. JLB looked immaculate – his best suit was crisply pressed – but as usual, he wore open-toed sandals, without socks. Even a date with the Prime Minister couldn't make J. L. B. Smith change his hard and fast rule of never wearing closed shoes. The Dakota touched down in Grahamstown to pick up Margaret Smith and William, then took off again, flying over Knysna, where JLB's eldest son, Bob, was spending the holidays. 'We had no idea what was happening,' Bob recalls. 'We had no radio and no newspapers. Then a plane flew over the house and Dad dropped a message for us out of the window, written on a board and attached to a parachute made from a sheet.'

As they were starting their approach into Cape Town, Captain Letley passed back another note: 'Dr Malan thanks you very much for having taken the trouble to come so far, but he does not wish to see the fish and wishes you a safe return to Grahamstown.' Smith was crestfallen. He had come to revere the Prime Minister, and was eager to show him the extraordinary product of his assistance. His first thought was that it must have been the evolution connection that had upset the Calvinist doctor. He shrugged his shoulders and said, 'OK, let's go to Cape Town anyway for lunch.' It was only when he saw the grins on the pilots' faces that he realised he'd been 'had' for the second time. The party landed at Cape Town, and while William was left at the Air Force base to look at the fighter planes, the Smiths and the coelacanth were driven in a military vehicle to see the Malans.

They were welcomed warmly. The coelacanth, still in its coffin – Smith had refused to show it to anyone before Malan had a chance to see it – was placed under a tree, and the lid opened for the Prime Minister. He looked at it, then turned to Smith and said, with a twinkle, 'My, it is ugly. Do you mean to say we once looked like that?' It was an extraordinary comment to come from a deeply religious man, an avowed believer in Adam and Eve.

After lunch, the fish was shown to a select group of people invited by the Malans. On his way back to the airbase, Smith was recognised by the taxi-driver: 'My God, sir, you aren't the genilman with the fish, are you? Oh, what an honour, what an honour for me and my taxi.'

Showing the second coelacanth to its namesake, Prime Minister D.
F. Malan.

The next day, on New Year's Eve, the Flying Fish Cart (as
it had been dubbed by the *Pretoria News*) took off on the last
leg of its epic journey. It made a short detour to circle over
Malan's house, to drop off copies of the morning papers, as
he stood on the lawn waving. A few hours later, the Dakota
arrived at last at Grahamstown, where a crowd was waiting to
greet it. Among the well-wishers were the Mayor, in full may-
oral regalia, and Marjorie Courtenay-Latimer, bursting with
pride. JLB appeared at the door of the plane, followed by his
family and, lastly, by the coffin bearing the coelacanth, which
was put on display. 'There was great excitement,' Marjorie
recounts. 'We were very thrilled about it, especially that it
appeared that they had found the home of the coelacanth in

the Comoro Islands.' Mrs Smith was given a bouquet and JLB presented a copy of his book to the four Air Force officers and waved them goodbye. The coelacanth was loaded with Miss Latimer into the museum's station-wagon and taken to J. L. B. Smith's laboratory at Rhodes.*

The safe arrival home of *Malania* was just the beginning for J. L. B. Smith. At last, he had a complete coelacanth to examine, the result of a fourteen-year search. He was inundated with requests for articles, interviews and 'viewings' of the fish. This time there was no war to overshadow the coelacanth story, and the world's press were free to turn the spotlight on to an ancient fish and its eccentric saviour. Despite JLB's anti-social tendencies, he revelled in the attention and was loath to deny interviews to even the most humble of newspapers and magazines. Although he had hardly slept for a week, he could not afford the time to recover: there was too much to be done.

He wrote an account for *The Times* of London, which was typed and sent on New Year's Day (Smith had called the Mayor at 5.30 a.m. to ask for two typists), and published the day after. In it, he tried to explain the importance of the coelacanth to mankind, and to science. 'It is a stern warning to scientists not to be too dogmatic,' he wrote. The fact that there appeared to be at least two species of a fish previously

*To commemorate the fortieth anniversary of the coelacanth flight, the Dakota, which was still in active service, was given a spit and a polish, and flown back to Grahamstown, with a full complement of passengers, including three of the original crew members: navigators Duncan Ralston and Willem Bergh, and radio operator 'Vanski' van Niekerk. Two years later, the Dakota was officially retired and donated by the South African Air Force to the SAAF Museum in Pretoria.

and confidently predicted to be dead, reveals the shallowness
of man's knowledge of what goes on in the oceans that cover
most of the earth's surface. 'We have in the past assumed that
we have mastery not only of the land but of the sea,' he wrote.
'We have not. Life goes on there just as it did from the beginning. Man's influence is as yet but a passing shadow. This discovery means that we may find other fishlike creatures
supposedly extinct still living in the sea.' In addition, Smith
stressed how the discovery of a living coelacanth had been of
immeasurable assistance to palaeontologists. Its remarkable
similarity to earlier reconstructions of coelacanths from fossil
remains had corroborated the accuracy of their attempts to
recreate long-dead creatures from faint imprints on ancient
rock. For the most part, he steered clear of the evolutionary
issue – particularly where it related directly to the coelacanth.
However, he emphasised that very little was known about the
internal workings of creatures from 400 million years ago, and
that an examination of *Malania*'s soft parts would teach us
something about how the internal organs of those ancient
creatures worked, enabling us to go a long way towards clear-
ing up the evolutionary picture. 'It may indeed prove to be a
sort of H. G. Wells's "Time Machine", only always in reverse,'
he wrote.

The Mayor of Grahamstown organised an official lunch for
local dignitaries, after which the coelacanth was put on display
to the public. Professor and Mrs Rennie, friends of the
Smiths's from Rhodes, were invited and remembered the day
vividly. 'It was a Saturday afternoon and the town was fairly
quiet otherwise, but there was just the biggest mass of people
I have ever seen waiting to go into the City Hall,' Bee Rennie
recalled. 'People were converging from all directions . . . From

judges to candlestick makers and heavens knows who. We saw the Judge President sort of pushing his way in, next to Helen Campbell the very short hairdresser.'

They described walking along a long passage to the hall, where the coelacanth lay in state in its coffin. Grahamstown's only traffic cop, Mr Archer, his eyes streaming from the formalin, directed the human traffic. 'It reminded us of King George V's death in 1936 when we had processed past the catafalque where his coffin was lying in state,' Professor Rennie recalled.

Margaret Smith also had vivid memories of the time following *Malania*'s return: 'It was the most exciting moment in a life full of interest. For days after returning to Grahamstown, we seemed to walk on air until we came down to earth with a bump when we realised how much there was still to do.'

Malania was later taken to the East London Museum, where it spent a few days with *Latimeria*, and to Port Elizabeth for similar dignified occasions. It attracted enormous attention wherever it went – around 20,000 people were estimated to have seen it in a short period of time. The Smiths received sackfuls of letters and telegrams from well-wishers from within the scientific community and without. One American ichthyologist wrote, 'Now I can die happy for I have lived to see the great American public excited about fish.' They appeared on television across the world, from America to Japan, Alaska and Timor. 'We were, in fact, carried along by a kind of tidal wave we could scarcely control, a wave that went right round the world many times, to its uttermost corners, and the backwash keeps on coming back to us even after this long time,' JLB wrote. 'In this process an obscure and highly technical scientific term became part of the common speech of mankind.'*

When Smith started his closer examination of the fish, he was shocked to note that the coelacanth's brain was missing, and much of the viscera had been badly damaged by its rough treatment. 'This was indeed a bitter blow. Once again it was not a complete fish,' he lamented. He was delighted to discover, however, many extraordinary and wonderful features, including gills that looked like jawbones, thus providing a clue that jaws and gill arches have the same origin. The hollow tube down its back he recognised as a notochord, a forerunner to a backbone and made of cartilage. He confirmed that the limb-like fins have their own internal skeleton, but there was no sign of lungs – instead, a swim bladder filled with fat. And he found, in its intestines, fish remains including scales and the eyeballs of a fish which Smith estimated to have been a good fifteen pounds in weight, confirming his belief that the coelacanth was a powerful and successful predator. In time, he realised that *Malania*'s missing dorsal fin and strange tail were probably a result of a shark attack, and not an indication that it was a different species from *Latimeria*.

The Smiths worked flat out on his account for *Nature*, which appeared in the 17 January issue – only three weeks after the heroic rescue. 'It is my great privilege to announce the discovery of a second Coelacanth,' he wrote, before thumbing a nose at E. I. White of the British Museum, and his theory that the coelacanth was a denizen of the deep. He

---

*A British MP, in attacking an opponent, called him a 'coelacanth' on the grounds that from his long silence in the House, he was surprised to find him still alive.

'If this is the best you can do in 50,000,000 years, throw me back.'

announced that he would welcome help from scientists in conducting the minute examination of his fish. Offers from specialists came flooding in, but in his heart, Smith did not want to chop up his *Malania*. While he knew that it would be in the best interests of science, he would have much preferred it if another, more complete specimen could be found for science, leaving *Malania* more or less intact.

There were others keen to have a coelacanth of their own. Early in 1953, the *Daily Mirror* reported that London Zoo was offering £1,000 for a live specimen. The Curator of Fishes, Mr Herbert Vinall, had apparently prepared a special large aquarium for it. 'We've had all sorts of queer fish in here,' Vinall said. 'I think we could make this old fossil feel quite at home.'

The 'crackpots' started writing again. Letters, mainly from religious people scattered across the world, castigated Smith for his work. One man published a denunciatory pamphlet about his views. There were threats of personal violence: Smith was told it would have been better for mankind had he never been born. He took these attacks with his habitual pinch of salt.

D. F. Malan could not escape castigation for his role in the seemingly apolitical act of saving an old fish. The *Manchester Guardian* printed a barbed editorial directed at the architect of apartheid, pointing out the disparity between his professed political and religious views, and his role in fetching the coelacanth. Under the title 'Daniel and Darwin', the leader read, 'It is not often that prime ministers resting from the cares of office are asked to interfere with their holidays in the interests of ichthyology. Dr Malan, it is reported, rose to the occasion with the disinterestedness of the true scientist. But did Dr Malan realise what he was doing? The Coelacanth is believed to be a link of evidence which might prove man's evolution from the fish stage. Dr Malan has been a minister of the Dutch Reformed Church, and only last September the Dutch Reformed Churches of South Africa complained of an exhibition at the Transvaal Museum which purported to trace man's evolution from the apes.

'The theory of apartheid and the superiority of White over

Black would take a nasty blow if it could be proved past doubt that all men, Black and White, have a common ancestor – and an even nastier tumble if it could be shown that the common ancestor is the Coelacanth, a mere fish.'

While Smith was grappling with the delicate inner parts of *Malania*, and basking in international fame, Eric Hunt, who had saved the second coelacanth for him, was having an altogether stickier time. He was proud to have been the agent of Smith's success, but he also foresaw problems about how the French were going to react to what they would inevitably regard as their humiliating failure to recognise the treasure on their own doorstep. He wrote to JLB from his boat the next day, describing lunch at the Governor's residence, during which Coudert had confided that he had already been on the receiving end of some 'blunt words' from the French Bureau of Scientific Research in Madagascar, about how he had stood by Hunt and allowed the coelacanth to go to South Africa. He believed that they had only held back from claiming the fish for themselves because they were sceptical of Hunt's identification.

Hunt described the ceremony in Mutsamudu in which the fisherman, Ahamadi Abdallah, was presented with his £100 award. According to custom, he shared a third part with his assistant (had the boat not been his own, he would have given a further third to the owner), but still, it was an enormous sum of money for a poor fisherman, and more importantly, the public reward conferred upon him a status above that of his peers. Ahamadi Abdallah told Hunt that yes, the fishermen knew of this fish. It was called *gombessa*, and was caught occasionally. *Gombessa* was not good when fresh, but could be

André Lehr presents the £100 reward to (from left to right) Ahamadi Abdallah, Soha and Affane Mohamed.

eaten salted. Abdallah confirmed that he had caught the *gombessa* at a depth of twenty to thirty metres, about 200 metres from shore.

Furthermore, Hunt reported to Smith, he had heard of another specimen, that had been caught recently near Mitsamiouli, on Grand Comoro, the largest of the Comoro Islands, but that had been thrown away when the local Muslim priest designated it unfit to eat. The scales were used, Hunt was told, for roughing up the inner tubes of bicycles when they got punctures.* And there were rumours of more

* Experiments by coelacanth guru Robin Stobbs, however, indicate that the scale is not strong enough for this to be the case. 'They might well roughen up a slab of frozen margarine but would be incapable of abrading a bicycle inner tube or of smoothing a piece of wood,' Stobbs maintains.

coelacanth catches, Hunt said, that he was inclined to believe.

He again thanked Smith for naming the fish after Anjouan. 'The Govt have always been very good to me here that I feel my insisting on Anjouan being included, was in a small way showing my appreciation of them. It is certainly a temptation to have such a rare discovery named after oneself, but in this particular case I felt my allegiance to the islands which have been my sole trading area for 4 years was stronger.' He asked Smith to send newspaper cuttings relating to the coelacanth to his father in London.

Two weeks after Smith and his fish had left the Comoros, Hunt's pride was followed by a bitter blow, when the *N'duwaro* was hit by a cyclone, and sunk. Hunt returned to Dzaoudzi to await the arrival of a 'Lloyds chap' to write off his boat. The storm had devastated the island, he reported. Dzaoudzi, 'which was in a way quite pretty, now looks as if an Atom bomb had hit it.' Houses had crumbled, and the streets were strewn with branches, rubbish and pieces of crumpled corrugated iron.

A French scientist was now installed on Anjouan, with orders to stay until he had got another coelacanth, Hunt reported in another letter to J. L. B. Smith. The Governor had come in for a considerable amount of criticism for allowing Smith to whisk the fish – caught by a French subject in French territorial waters – from under French noses. Long and furious articles had been published in Paris newspapers, railing against the '*vol*' of the stolen coelacanth. 'Mind you I can quite sympathise with them and see their point of view,' Hunt wrote. 'I am quite sure you would be furious if a French schooner type such as myself had got one in SA and sent it to France.'

Hunt's letter crossed with one from Smith, sent to Zanzibar, in which he described the events following his return to South Africa. He wrote of the broadcast, which had been played across the world by the BBC. 'I hope you heard it,' Smith wrote, 'because it paid full tribute to your ingenuity and determination ... [a] Mrs Waddington wrote to say she had heard my broadcast in England and she writes, "We are so glad that it was Eric. If you wanted a Dinosaur he would get you one."'

He suggested that Hunt might now find it more profitable to act as a coelacanth guide for all the expeditions that were bound to converge on the islands, than to continue hard trading. Indeed, Smith wrote, he and Margaret were most keen to revisit the Comoros later in the year, to investigate the 'great treasure house of fish-life'.

'Science is grateful to you for your part in this important discovery, and I hope you will use your knowledge to promote others,' Smith concluded. Margaret Smith sent her own letter, in which she congratulated Hunt: 'As long as the human race continues, your name will be linked with one of the most thrilling stories of our age. It is very gratifying to me that my confidence in you was so well founded and that you were in no way to be overawed ... You have certainly attained one of your great ideals: to put the Comoro Islands on the map. After this we expect them to receive a considerable amount of attention from the outside world.'

In France, the anti-Smith sentiment was hotting up, and unfortunately for Hunt, stranded in the Comoros after losing his ship, he was well placed to take the flak. Pressure was put on Governor Coudert to do something. It was he, after all, who had allowed Smith to whisk the coelacanth away from French territory, and he was now forced to take action to save his career. He turned on the nearest scapegoat and proceeded to confront Hunt and to write to Smith contradicting Hunt's version of events, and presenting a very different scenario.

At this point, Hunt's gratitude towards the French started rapidly to fade. Obviously upset, he wrote to Smith, 'I was furious and told the Governor that over the whole affair I had not sought to make a personal gain out of it, and therefore the least I could expect was the credit due to me for having started the necessary propaganda, and personally guaranteed the payment of £100. Also, having been the only person concerned with the preservation of the fish. <u>Nobody bar myself and my crew who took turns with me in injecting the fish, had anything whatsoever to do with it.</u>'

Hunt laid out in detail the steps he had taken to secure the coelacanth, and the Governor's co-operation in the process. 'Paris was furious that the fish had not gone there. You say ultimately it would have come back to you. I am not in a position to judge, I don't know. But I do know this, that Paulion who I talked with in Tananarive was furious with the Governor . . . [who] in order to extricate himself a little wrote his letter to you cutting me out of the affair almost entirely, and more or less making me out to be only the transport for the fish from island to island. Claiming that he had handed the fish over to you in the name of France and that in Anjouan the Govt had taken charge of the fish and treated it before you came into the picture.

'I am really quite upset over all this, as from being a friendly affair between ourselves, it has turned into quite a political one,' Hunt wrote.

He explained that the French had made him sign an official version of events which differed radically from the true story. There were three versions, Hunt explained: the official story sent to France before Smith's arrival, which corroborated Hunt's account; the doctored version in response to pressure from Paris, which effectively cut Hunt out of the story and which he refused to sign; and the compromise version, which still put a heavy French spin on events. 'The Comoros are my means of earning my living,' Hunt wrote, 'and studying fish, and fishing my hobby. I must by this reason remain on good terms with the French authorities, and to that end I have signed the latest official story of the Coelacanth ... I wish now that I had accepted your offer of naming the fish after me, it would certainly have guaranteed me a certain amount of credit, which I have not received.

'Without being too much of an egotist, I think that you appreciate that if I had not met your wife in Zanzibar there would have been no 1952 Coelacanth for anybody.'

Smith replied in lofty tones, reassuring Hunt that while he was 'obsessed with what those Authorities have said, this is not what has been published all over the world. You need not feel you have not been given ample credit. You have ... You have actually had more credit for asking that the fish should not be named after you, but after Anjouan ... Both my wife and I have seen that your ability was appreciated ... Mrs Smith has repeatedly said in public that she always felt that you were the most likely layman to be able to spot an animal of that kind.'

He played down the actions of the French, suggesting that they did not 'really understand the full position. They were so astounded at the world's reaction to this that they felt they had been deprived of something important,' he wrote. 'What they did not realise was that if anyone else had got another Coelacanth, it would certainly have been interesting, but the terrific reaction everywhere was not by any means the Coelacanth alone, but of the remarkable circumstances surrounding it. First of all perhaps most interesting to the world at large was the fact that this was the direct result and culmination of my fourteen years' research. Secondly, there was your part in it in direct relation to us and to the long search.' Malan's intervention, too, accounted for some of the interest, Smith conceded. 'None of these things would have been attached to the discovery by any other person. Furthermore, had the French seized the specimen the world would have known that they had got it not by their own efforts, and they would certainly not have been admired for their action.'

Smith apologised that he was not in a position to reimburse Hunt for outstanding expenses – the telegraphs and time lost in waiting for the plane – over and above the £100 he had already given him, but added that he hoped to raise funds for an expedition to the Comoros later in the year. Sadly, that was not to come about. The Smiths and Hunt stayed in contact with each other, but they were never to meet again.

Hunt's involvement in the coelacanth episode – notwithstanding the abuse he had had to take from the French for J. L. B. Smith's dramatic appropriation of what they saw as a French fish – further stimulated his interest in the aquatic world. After the *N'duwaro* was wrecked in 1953, he bought a new, larger, 100-foot schooner in Naples, which he named *Hiariako* (Swahili for 'your choice'). Two years later, he married Jean Fowler, a Scottish girl fourteen years his junior, in

Tananarive, the exotic French colonial capital of Madagascar. She frequently accompanied him on his voyages, but Hunt still found the life of an ocean trader incompatible with that of a husband. Fish had, by this time, evolved from an interest into a passion. He not only kept an aquarium at home and on his ship, but had started to study them in detail, and to become involved in the conservation of the marine environment. He planned to sell his ship, and settle in Madagascar with Jean, and to set up a business collecting exotic fishes for aquariums around the world.

On 9 April 1956, the Hunts set off from Zanzibar, in search of fish specimens and cargo. Eric had confided to a friend before he left that it would probably be his last cruise before he sold up. They spent a few days in the Comoros, where they dropped off some cargo, then sailed for Majunga, a pretty town on the north-west coast of Madagascar. There the Hunts collected fishes, which they transferred into the on-board aquarium. Jean then left, to search for Madagascan lace plants and visit friends in Tananarive, while Eric headed back to the Comoros with another load of cargo. He sent her a telegram telling her to fly over and join him in Dzaoudzi, where he would meet her at 9.30 on the morning of 3 May. He never arrived.

According to an official inquiry, the *Hiariako* set sail from Majunga on schedule in the early hours of 1 May. With Eric were fourteen crew and eleven passengers. Before long, they found themselves being jostled by large waves and swept along by strong winds. By the next morning, Hunt had to admit he was lost. All that day and night, they searched in vain for a sighting of land, and at 4 a.m., while Eric was asleep, the vessel struck the Geyser Reef, and slowly ground to a listing halt on the treacherous coral, only eighty nautical miles from their destination.

The crew tried desperately to re-float their ship. They jettisoned all the cargo, but water continued to pour into the unfortunate *Hiariako* and Hunt was forced to give orders to abandon ship. He, a French passenger and the cook, climbed into a small sailing dinghy; the senior members of the crew, with three Comoran women and three children, took the ship's boat; while the remainder spread themselves across a wooden raft and another makeshift raft constructed of oil drums and hatch covers. Each had a little fresh food and water.

At first, Hunt attempted to tow the other vessels to shore with his small outboard engine, but he soon realised that wasn't going to work. He decided to hoist sail and set off with the Frenchman and the cook to try to raise help from Mayotte. He was never seen again. A vicious storm engulfed the sea, and probably took the people on the rafts with it. The ship's boat managed to out-weather the storm, but the nine people on board soon ran out of food and water, and within days, two of the children and one of the Comoran women had died. Fifteen days later, on 20 May, the survivors were spotted by Comoran fishermen in three small outrigger canoes, and taken back to the safety of Grand Comoro.

On 24 May, Hunt's dinghy was found, on its side and empty, fifteen miles west of Moheli, the smallest of the four Comoran islands. The engine appeared to have been torn away from its bracket, and there were no signs of its human occupants, apart from the floating remains of two Mae West life-jackets some 300 feet away. The bodies of Hunt and his two companions were never recovered.

The shipwreck caused excitement in the British press, when an erroneous report in the *Daily Telegraph*, quoting unnamed British Intelligence sources, suggested that Hunt had been killed during a daring attempt to spirit the leader of the Greek Community in Cyprus, Archbishop Makarios, away from his Seychelles island exile. Hunt's family were terribly upset by this slur on Eric's character, and numerous angry letters were exchanged with an unyielding *Daily Telegraph*. The official report into the accident, however, makes no mention of kidnapping missions, and concludes that it was an accident, for which neither the captain nor the crew were to blame.

In an obituary titled 'An Aquarist Dies at Sea', Henry Nichols FAI wrote, 'Eric Hunt will be missed deeply by all his fisherman friends, far and near; but also his departure is inevitably going to leave a very empty spot in the lives of simple, humble black and brown people along the central East African coast from Pango to Lourenço Marques and on many of the offshore Indian Ocean islands ...' Nichols wrote of how Hunt helped the locals with money and goods as well as his friendship and trust. 'He was an artist,' he concluded, 'in the best sense of the word. A man who lived the life he loved and loved the life he lived till its last roaring, foam-swept moment – trying to help somebody.'

J. L. B. Smith was left bereft, when in 1953 the French refused him permission to return to the Comoros. 'The global reactions to the extraordinary culmination of my long search for the Coelacanth had profound effects in France, where widespread and somewhat hysterical propaganda in the press aroused public feeling,' he wrote. 'As a result, there was widespread agitation, and it was urged that the French

Government should demand the surrender of the Coelacanth to France.' A diplomatic incident was brewing: the loss of *Malania* clearly still rankled and so it was no surprise when on 9 November 1953, Paris issued an edict: 'Only French scientists will be allowed to search for the coelacanth off the French Comoro Islands, in the Indian Ocean between Mozambique and Madagascar, for the rest of this year. French authorities there have declared a complete ban . . . on expeditions by foreign scientists . . .'

The coelacanth had become a French fish.

VI

NOTRE COELACANTHE

The French were determined to make up for lost time, and to claim the coelacanth as their own. They knew they had initially underestimated the excitement that *Malania*'s discovery generated. Both the scientific world and the public at large were ablaze with interest in the strange and ancient fish, and the French wanted to be a part of it. Even Hollywood latched on: *The Creature from the Black Lagoon*, featuring a finned monster that emerged from the sea, was apparently inspired by the discovery of *Malania*.

Dr Jacques Millot, a spider expert with an uncommon resemblance to a kindly hobbit, at that time headed up the French research facility in Tananarive, Madagascar. He took charge of co-ordinating the search for a French coelacanth, and for the next two decades, foreign scientists continued to be excluded. The glory of unravelling the intricate inside workings of the world's most celebrated fish was to be French alone.

Millot installed one of his colleagues, Pierre Fourmanoir, in the Comoros with the aim of instigating a concentrated fishing effort among the local fishermen. Needless to say, the £100 reward they were promised for what had become known as '*Le Poisson*', was sufficient spur to galvanise their efforts.

On 26 September 1953, a 51-inch (1.29-metre), 87-pound (39.5-kilogram) coelacanth was caught off Mutsamudu, Anjouan, the far side of the island from where *Malania* was fished up. The fisherman, Houmadi Hassani, recognised it immediately, and quickly took it back to his house. His wife stood guard while he ran for help to the French doctor, Dr Georges Garrouste, who had been equipped by Millot with a special Coelacanth Preservation Kit. He collected the fish in his ambulance and injected it with seven gallons of formaldehyde. The next day it was flown by a special aircraft to Millot in Tananarive. The French were overjoyed. Pictures of the specimen adorned the front page of *Le Monde* under the jubilant headline: '*Notre Coelacanthe!*'.

J. L. B. Smith was on the island of Bazaruto, off the coast of Mozambique, at the time. He was in the process of examining a huge rock cod, which had been found in very shallow water, when he was approached by a Portuguese-speaking Chinese man. This man had heard a report on the radio the previous night about a large fish with a strange name that had been caught by the French, somewhere near Madagascar, he thought. Smith's name had been mentioned, which was why he had approached him. Smith interrogated the man, and established that it was definitely a coelacanth, but beyond that he could elicit no further information. He searched for a recent newspaper, or someone else with a wireless who had heard the broadcast, but it was not until he got back to civilisation a week later, that the report was confirmed: the French had caught their own coelacanth in the Comoros.

'I shall always remember the sensation of terrific relief this gave me, as if a crushing burden had been lifted from my mind,' he wrote in *Old Fourlegs*. 'So it was the right spot, they were there! I could see the end of this great strain. I could keep my *Malania*, and it would be only a matter of time now before all those high specialists were each and all wresting the secrets of the life of the long ages past from the tissues and structures of fresh coelacanths.' Smith immediately sent a telegram of congratulations to Millot.

The small volcanic islands at the western edge of the Indian Ocean were, in Smith's mind and the minds of most people, confirmed as the home of the coelacanth: the aim of his fourteen-year search had been fulfilled. What nobody considered at the time, was that the Comoros are relatively young islands: the oldest, Mayotte, was pushed up from the sea only five and a half million years ago; Grande Comore, a mere 130,000 years ago.

The year 1954 brought a bumper harvest of coelacanths for the French. Their second specimen – the fourth to have been officially recorded – was taken near the village of Iconi in Grande Comore, indicating that coelacanths lived off at least two of the four islands that make up the Comoran archipelago. Its capture, on 29 January, was followed by a similar dash to box and transport it. 'It was very exciting, rushing to preserve it, to build a case, to order a special plane from Madagascar,' noted an American reporter from *Colliers* magazine. 'We finished up tired and proud at 4 p.m., then a man staggered in with an even bigger coelacanth,' recalled Maurice Rex, the administrator of Grande Comore. Specimen Five, caught near the small village of Mandzissani, was boxed up and sent with Specimen Four to Millot in Madagascar.

Suddenly there was no shortage of coelacanths to dissect. Up to this point, no photographs had existed of coelacanths that were not damaged and decomposing, or stuffed, rigid and brown. For the first time, Millot saw what Marjorie Courtenay-Latimer had described as 'the most beautiful fish'. He and his assistant, Jean Anthony (a shy comparative anatomist), were soon elbow deep in coelacanth viscera. They started work on what would become the three-volume, profusely illustrated *L'Anatomie de Latimeria*, the definitive study of the anatomy of the coelacanth, and probably one of the most detailed ever examinations of a fish, which would take them eighteen years to complete.

Much had already been written about the coelacanth's external features by the time Millot got to work, mainly by J. L. B. Smith in his detailed monograph on *Latimeria*. The world was familiar with its hard, mottled scales, its wide mouth with those generous lips, its large eyes. Close attention had been paid to the skeleton of its limb-like fins, which are unmistakably similar to early tetrapod limbs, and to its unusual fringed tail, seemingly a continuation of the body, with its small 'puppy dog' extension. Now, with the benefit of a complete coelacanth, Millot and his team were able to discover the detailed inner workings of this relic of ancient times.

Several features exhibited marked similarities to those of living tetrapods: the coelacanth's inner ear was found to have more features in common with frogs than with other fishes; its gills were similar in shape to jaws, with gill rakers resembling teeth, and its blood was found to contain large red cells, similar to that of amphibians and lungfish. All of these confirmed that the coelacanth was indeed very closely related to the first fish that climbed out of the water to found the greater part of the animal kingdom on land.

But there also emerged information that for a short time appeared to indicate a kinship with sharks, among the most primitive of fishes. The coelacanth's blood, for instance, was also found to be rich in urea, like that of sharks. Instead of a backbone or ribs, it has a cartilage notochord — hollow spinal column — similar in composition to the cartilaginous discs of sharks. The coelacanth's notochord is filled with a unique oily fluid. Its V-shaped, symmetrical heart is also like both a shark's and that of most vertebrate embryos, and Millot found a spiral valve in the coelacanth's large intestine like that found in sharks, which allows them to digest their food very slowly and completely, and to go for longer between meals. This is clearly an adaptation to living in food-poor areas.

Indeed, that the coelacanth appears to be singularly well suited to its environment is not really surprising in such a successful survivor. It has a very slow metabolism, which means it can conserve energy when food is scarce. It also has a small gill area — less than that of any fish of comparable size, and one hundredth of that of a tuna — so that it absorbs oxygen at a slow rate, making it well adapted to living in deeper, cooler areas, where the water has a higher concentration of oxygen, and where there are fewer predators and less competition for available prey. The coelacanth's slender, tubular swim bladder is filled with fat and other non-compressible body fluids (unlike the hollow, gas-filled organs of bony fishes), which gives it neutral buoyancy, allowing it to hover comfortably at great depths, without having to move and thus expend energy. It has a reflecting layer behind the retina of its eyes which, as in cat's eyes, gives it good vision in dim light (and conversely, as the eye contains no melanin, blinds it in bright light). The Comoran fishermen describe coelacanth eyes shining like 'lights or burning coals'.

The coelacanth has a pair of features thought to be unique among living creatures. Its brain – which is primordially simple and very small, weighing about $1/15{,}000$ of its adult body weight and occupying less than $1/150$ of the brain cavity, about the same size as a grape – lies behind a hinge-like intracranial joint which separates the nasal organs and eye from the ear and the brain. This joint is considered to be very primitive, and is thought to allow the fish to open its jaws wide, and thus increase the power of its bite. Similarly unique are the six jelly-filled cavities embedded in the snout. Scientists have christened this feature the 'rostral organ', and speculate that it is a highly complex electro-detection system, used to locate prey.

The inner workings of the coelacanth were like a series of signposts pointing in different directions. Part of its appeal was undoubtedly its well-publicised propinquity to the main trunk of the evolutionary tree. Millot and his team had clearly hoped that the availability of a modern descendant to study would help to clear up the ongoing and often fierce debate as to which of the lobe-finned *Sarcopterygii* was the first to crawl out of the water and colonise the land. Their early results, however, were leading them around in circles, and after a few years, the missing link between fish and land-dwelling creatures had still to be confirmed.

The lungfish had been among the earliest contenders. The first living specimen was discovered in the Amazon by a

Viennese naturalist called Johann Natterer. He returned to Europe in 1836 from eighteen years of travel in Brazil with an enormous collection of exotic creatures that increased the inventory of the Imperial Museum in Vienna six-fold. His prize discovery was a fish-like beast, about two feet long and eel shaped, with both gills and fully functioning lungs. The locals called it *caramuru*, and Natterer coined the scientific name *Lepidosiren*. He wrote in a monograph published shortly after his return, that it was a 'new species of animal of the family of fish-like reptiles (Ichthyodea). It deviates in all its characteristic details so significantly from other representatives of that group [of reptiles], and resembles a fish so closely in its overall formation that even the most experienced natural investigator can be misled.' Time would prove that it was Natterer himself who was thus fooled.

One year later, a similar lungfish was brought back from the Gambia River in West Africa by Englishman Thomas Weir, along with a second, dried specimen enclosed in sun-baked clay. This lungfish was known locally as *comtok* and christened scientifically as *Protopterus*. It was soon discovered that, during the dry season, both species curl up and go into aestivation in round mud-cakes, and only reawaken when the rains come and soften their muddy nests.

Unlike coelacanths, lungfish fossils at that time were not widely known, so the arrival on the scene of living creatures was regarded less as a surprising comeback than a startling bolt from the blue. Natterer's initial monograph had sparked a fierce debate among scientists as to what exactly they were: fish or amphibian? The question, according to one scientist, had 'singularly embarrassed taxonomists'. Under most classifications, fish had scales and gills, while amphibians – with naked skin – possessed the ability to breathe. What did that then make *Protopterus* and *Lepidosiren*, with their scales, gills *and* lungs?

Contrary to Natterer's initial diagnosis, majority opinion settled on the side of them being breathing fish (rather than scaly amphibians), possibly somehow related to the first tetrapods. The problem was that they were clearly not walking fish, as neither had fins to speak of – just thin, string-like appendages – and they swam in an eel-like manner, their whole body squirming. Scientists, led by Darwin and Wallace, were convinced that the true stem of land-dwelling creatures possessed both the ability to breathe air and the legs to walk on land. The lungfish fulfilled their first criterion, but appeared to fail miserably on the second. Indeed, up until that time (before 1839 when the first coelacanth fossils were unearthed), they still hadn't found any fossil evidence of a fish with limb-like fins. To fill this 'missing link', the anatomist, Karl Gegenbaur, had invented such a fin with 'leg possibilities' – his students christened it *Archipterygium gegenbauri* (Gegenbaur's archaic flipper) – evidence of which, he prophesied, would some day be found in a fish-like creature that could breathe air.

Then in 1869, an Australian bushwhacker called William Forster moved to Sydney from his cousin's farm near the Burnett River in Queensland. Soon after arriving in the city, he visited the Sydney Museum, where he fell into conversation with the curator, Gerard Krefft, about the unusual animals found in Australia. Forster asked why the museum had no example of a strange fish that lived in the Burnett River. This fish, Forster explained, was known locally as the Burnett salmon. Krefft was interested and asked him to describe the fish. It was about five feet long, Forster said, round like a fat eel, with large greenish scales and four strong fins. Krefft said he had never heard of a fish that fitted that description, but would very much like to see one. Forster promised to write to his cousin, asking him to send a Burnett salmon to the museum.

A few weeks later, a large barrel arrived for Krefft containing several specimens, strongly salted to prevent them from decaying. He took out a five-foot-long fish to examine. Forster's description had been pretty accurate: Krefft saw the scales, and the four unusual fins which looked more like paddles. He was most struck by the tail, which was unlike any fish tail he knew: it looked more like an extension of the body, fringed with fin material. But it was when he carefully prised open the fish's mouth to look at its teeth that he was stunned. He immediately recognised the four large teeth, which were joined together like a rooster's comb, to be remarkably similar to fossil remains of teeth found in ancient rock deposits. No living fish he knew of had teeth like that. Nobody had worked out what the ancient creature belonging to those teeth had looked like; it had even been suggested, by the palaeontologist James Parkinson, that they represented the 'digitated termination of the sternal plate' of a tortoise. Louis Agassiz – the same Agassiz who had named the first coelacanth fossils – had christened the owner of these teeth *Ceratodus* (the horn-toothed).

Krefft delved into the archives to search for further clues, and was astounded at what he found. Nowhere was there mention of a living fish with fins matching the description of the Burnett salmon's; the only similar sort of fins he could find were Gegenbaur's imaginary archaic flippers. Could this fish lying in front of him on the dissecting table possibly be the missing link, and not a mere fossil, but the real thing? Krefft was filled with excitement. He continued his dissection and was overjoyed to discover that the so-called salmon had, as well as gills, a single lung. He came to believe that he had stumbled upon the original primitive lungfish, a forefather of *Protopterus* and *Lepidosiren* and, quite possibly, of all mankind. He christened it *Ceratodus forsteri*. When a complete fossil was later found which matched exactly Agassiz's *Ceratodus*, and which was subtly different from the modern Burnett salmon, the name was changed to *Neoceratodus* (new *Ceratodus*).

For a long time, *Neoceratodus* was a zoological sensation. It was to the nineteenth century, what *Latimeria* is to the twentieth, and it was as exhaustively dissected, examined and debated. Ernst Haeckel, who succeeded Darwin as the 'greatest living zoologist', took a great interest in the Australian lungfish, and became determined to investigate its ontogenesis. Ontogeny was at that time a fashionable theory: it stated that during the time it takes to develop from fertilisation to birth, an embryo hustles through all the forms of its ancestors, in a rapid, compressed manner – rather in the way that the life of a drowning person flashes through his or her mind. If this is true – and recently the theory has lost much credence – the study of the development of any embryo can be used to explain much about its ancestry. A human egg, for example, would start off with primitive fish-like characteristics, and throughout the nine months of gestation, would resemble, in turn, an amphibian, a reptile and a monkey, before taking on the sophisticated senses of a human.

Haeckel, an impressive, erudite and multi-honoured man (by the end of the nineteenth century, he was known formally as His Excellency, Herr Geheimerat Professor Dr Ernst von Haeckel), was obsessed with ontogenesis. For decades, all European schoolchildren were made to repeat his famous formulation: 'Ontogeny recapitulates phylogeny' – the development of the individual mimics the development of the species – until it was etched indelibly on their brains. He dispatched one of his disciples, Professor Richard Semon, to Australia to look into the ontogeny of *Neoceratodus*. Semon's plan was to get hold of hundreds of lungfish eggs and monitor them minutely throughout their development process. He arrived in 1891, and disappeared up the Burnett River to a remote spot, where he pitched camp and hired Aboriginal egg-finders – who knew of the fish as *dyelleh*. He soon came upon a large, female lungfish, and settled down to wait for her to spawn. He had learnt that the fish laid its eggs on the stems of aquatic plants, so when the first eggs were found, Semon sent his team out to investigate every plant in the river. For a few days, they brought in eggs. He carefully labelled each one, and at regular intervals, broke open an egg, and preserved it in alcohol. His aim was to produce a full pickled spectrum of embryonic development. After a short time, however, his supply dried up. No more eggs were to be found anywhere. Semon quickly discovered the reason: despite being strictly forbidden by him, the Aborigines had caught the female lungfish and eaten her. He realised with horror that he would have to wait a full year until the next spawning season, and then brave the difficult conditions again.

The following year, Semon was more successful. He hired different Aborigines, who proved to be more obedient, especially after he offered rewards of $25 for the first eggs to be brought in. That year, 700 eggs were collected, appropriately labelled and pickled. Semon went back to Germany to work on his observations, which he collected in an exhaustive paper. He had learnt, among other things, that *Neoceratodus* does not aestivate (entomb itself in mud during the summer) like its African and South American cousins, and that its single lung cannot keep it alive on dry land. He also discovered, however, that as long as there is sufficient moisture, it has – rather like its cousins, *Lepidosiren* and *Protopterus* – internal nostrils which help to keep it going during the dry season, when other fishes perish. Even when in plentiful fresh water, he observed, the fish came to the surface for a gulp of air about every hour.

As for its strange, limb-like fins: while *Neoceratodus* often used them like legs in the water, they were not strong enough to support it on land. Semon's most important conclusion of his work into the ontology of the Australian lungfish was that, contrary to what had previously been believed, it was not a direct ancestor of amphibians – more like a close cousin. The race to identify the first walking fish was still on.

Determining the interrelationships of creatures is a complex process and scientists are forever inventing new ways of doing it. In his book *Living Fossil – The Story of the Coelacanth*, biologist Keith Thomson describes a conundrum put before a meeting of zoologists in Reading. Which of the following three creatures, they were asked, are most closely related: the salmon, the lungfish and the cow? (They might easily have used the coelacanth instead of the lungfish.) The answer to non-scientists might seem immediately obvious: the two fishes. But not to zoologists. To them it is clear that it is the lungfish and the cow, because they belong on a similar line of descent and thus have a more recent common ancestor: the lungfish is at least a cousin of tetrapods (albeit many times removed), while the salmon, whose line branched off much earlier, is only a cousin of the whole group. Thomson uses an analogy: who are most closely related out of Queen Elizabeth II, Kaiser Wilhelm II and Margaret Thatcher? Although the Queen and the former Prime Minister are both female, British, late-twentieth-century rulers of countries, the Queen is clearly more closely related to the Kaiser, since both are descendants of Queen Victoria. Thomson's point illustrates that in deciphering relationships between organisms, one must not look merely at appearances, but at ancestry and the clues provided by uniquely shared characteristics. But the characteristics of the contenders are complex and often conflicting.

An examination of the shared characteristics of the Australian lungfish, the living coelacanth, a living amphibian, say a newt, and a common ray-finned fish such as a salmon, showed clearly that the lungfish had most in common with the newt, while the ray-fin had least. The coelacanth was somewhere in the middle – exhibiting enough newt-like characteristics, like its limbs, inner ear and gills, to show that a relationship existed, but not enough to indicate that it was a very close one.

The three ancestral lobe-finned fishes, the fossil coelacanths, the now-extinct rhipidistians, and the primordial lungfish, shared characteristics – with each other, and with the fossils of early amphibians such as *Ichthyostega*. Of particular notice was the structure of the head casing and paired fins. From fossil reconstructions, one can see that a rhipidistian looked like a longer, thinner, smaller version of today's coelacanth; with the same strong notochord and pudgy fins. Fossil evidence of the Rhipidistia died out in the early Permian period, around 300 million years ago (although, after the surprise 'reappearance' of the coelacanth, also thought to have been extinct for many millions of years, one has to be careful in signing its formal death certificate).

Since the fossils of all of these creatures came to light in the nineteenth century, scientists have been debating the inter-relationships between the different lobe-fins and fighting for their particular favourite to be acclaimed as the true ancestor of man. For a while, in the first half of the twentieth century, the coelacanth was flavour of the month; then the accolade passed back to the lungfish, on to the Rhipidistia and back again. Even today, more than sixty years after the discovery of the first coelacanth, there is still no clear conclusion. The debate looks set to run and run. The latest theory is that, while the lungfish is a 'sister' of the first tetrapod, the coelacanth is a 'cousin', but again this has not been proved. Even direct comparisons of the genetic codes of the modern descendants through DNA analysis have been inconclusive. Recent results show that coelacanth chromosomes are remarkably similar to those of primitive frogs, and that the lobe-fins and tetrapods are clearly related; however, within the group, none show up as being more closely related than the others.

DNA testing was not available to Millot and Anthony in the 1950s and 1960s. However much they learnt from their investigations, much about the coelacanth remained a mystery. Quite apart from complex issues like where exactly the coelacanth fitted into the evolutionary tree, they still couldn't answer more fundamental questions about the fish, such as

A question of prehistoric paternity: 400,000,000 years ago, one of these three fish crawled out of the sea to conquer the land.

how it reproduced. One fossil showed the vague imprints of tiny babies inside the outline of the adult. This was read as evidence that coelacanths gave birth to live young – fully functioning little coelacanths rather than eggs. (This indicates a high level of sophistication – only five per cent of fishes are, like mammals, live-bearers: a female cod, for instance, lays a million eggs, which are swept away by the current, and relatively few of which get fertilised, while lungfish, like most amphibians, lay large, yolky eggs into nests, which they protect from predators.) In 1955, and again five years later, this theory was apparently thrown out of the water when the first female coelacanths were dissected by Millot and found, like sharks again, to carry immense, delicate eggs – the size of grapefruits, the largest known fish eggs. Perhaps the coelacanth was not, after all, ovoviviparous – live-bearing – but oviparous, egg-bearing?

After years of detailed study, Millot and Anthony had gathered an enormous amount of information on the coelacanth. However, rather than supplying simple answers, their work raised more questions. The coelacanth was not simply a key to the past, and the door to its ancestry could not immediately be unlocked by an examination of its anatomy. It was clear that, in *Latimeria*, they had a fish with an unique combination of characteristics, some of them in common with amphibians, some with fish, and which, over millions of years, had found different solutions to the problems posed by evolution. What was as interesting as what it shared with ancient creatures, was how and why it had survived for so long when so many other creatures had perished: why it had endured to become a living fossil.

When Millot and his colleagues opened the time capsule

and revealed to the world *Latimeria*'s inner secrets, they did not find a blueprint for our ancient ancestor, but instead a confusing pot-pourri of characteristics. They soon came to accept that they were dealing not with an old fossil, but with a contemporary fish, which had adapted itself in order to survive for 400 million years while the world around it – the oceans and the continents – had changed beyond all recognition. Few of its fellow creatures had been around for much longer than a twinkling in the coelacanth's eye – even the Indian Ocean had only come into existence more than 100 million years after the coelacanth first swam in primordial waters. The fish on the slab in front of Millot was interesting as more than just an obscure link to our distant past.

VII

LE POISSON VIVANT

Fairly early on in his investigations, Millot had realised that if he was to make any major progress in understanding why and how the coelacanth had survived for so long, while the vast majority of creatures had become extinct, he would have to see a coelacanth in its natural environment. It was one thing to note the limb-like fins, or study the muscles – but it would be quite another to see how they were used. It was imperative for science to get hold of a live coelacanth.

This was not to be an easy prospect. In 1954 the famous Jacques Cousteau was summoned to the islands in his equally well-known ship, *Calypso*, to try out his pioneering deep-sea diving methods in Comoran waters. Under Millot's direction, he took reels of submarine photographs in areas thought to be possible coelacanth lairs. But on this occasion, even Cousteau's legendary Gallic charm failed to entice the coelacanth from its ocean hiding place.*

* Cousteau tried again in 1963 and 1968 – and again he failed to find a coelacanth to film.

Millot turned his attention to the local fishermen. Comorans fish in narrow, banana-shaped canoes, each roughly hewn from the trunk of a single kapok tree. Each *galawa* is not much larger than a coelacanth, and hitherto, once a fisherman had caught a coelacanth, he would immediately batter it to death, or slit its throat to prevent it from struggling, and then haul it into the boat. The fishermen were afraid that if they left their valuable catch in the water, and attempted to tow it to land behind the boat, a shark or barracuda might grab it, and they would lose both the fish and their reward. So, in order to make the capture of a live coelacanth more attractive, Millot doubled the reward. He also set about devising a suitable watery jail for the captured fish.

The large reward was enough of an incentive for the fishermen, even if they were not altogether successful. Early one morning, Dr Guy Arzel, who was at that time in charge of the simple hospital on Grande Comore, was awoken by a pounding on his door. He opened it to find a fisherman cradling a bleeding coelacanth in his arms. He pleaded with the doctor to save its life: he had caught the fish only a short time ago, he explained, and when he saw it appearing out of the blue depths and realised that it was a *gombessa*, he was determined to bring it to shore alive and claim the large reward. As the fish broke the surface, however, his courage failed him, and in a fit of terror, he bludgeoned the fish repeatedly on its head. He had brought it to the hospital in the hope that Dr Arzel could somehow bring it back to life. Unfortunately, this was beyond the powers of the French doctor, who placed the fish on his operating table but, not knowing how to proceed, had to stand by and watch the beautiful blue creature before him slowly fade and die.

Shortly after this, an Anjouannais fisherman had better luck. On 12 November 1954, at 8 p.m., Zema ben Said caught the eighth coelacanth, 1,000 metres off Mutsamudu, on the north-western coast of Anjouan. It was two days after the full moon, and the sea was calm. From the way it took the bait, he guessed it was a *gombessa*, so he took his time, carefully hauling the fish in by hand from a depth of 140 fathoms (255 metres). Having made sure it was indeed '*Le Poisson*', Zema decided to try for the doubled reward. He gently passed a cord through its mouth and out of the gill opening, and thus he led the creature back to Mutsamudu jetty; though sometimes it was the fish that towed the *galawa*.

The administrator of the island was immediately notified, and as planned, sunk a seven-metre whaler off the end of the jetty, leaving the plug out of the bottom, to allow a small current to flow through. By 9.30 p.m., the makeshift aquarium was ready. The live coelacanth was released into it, and the boat was covered by a net to prevent the fish from escaping. Every half hour, the boat was rocked, to enable fresh water to enter. Witnesses reported a dark greyish-blue fish, the colour of watch-spring steel, with luminescent greeny-yellow eyes.

That night, the population of Mutsamudu celebrated the valuable catch. They sang and danced until daybreak, making regular trips to view their precious fish. The coelacanth, at first, appeared bewildered but otherwise calm; it swam slowly, by curious rotating movements of its pectoral fins, using its second dorsal fin, anal fin and tail as a rudder. As the sun rose, however, it became increasingly distressed, apparently reacting against the light and the heat, trying to hide in the darkest corners of its makeshift cage. Several tent canvasses were placed over the boat, in an attempt to protect the fish from the sun, but to no avail. By 3.30 p.m., it was lying belly up, its fins and gill covers making agonised movements.

Millot arrived from Tananarive on a special flight, just in time to see the coelacanth's last breaths. He hauled it from the water, covered it with a sheet and took it to the hospital, where it was formally pronounced dead – but in perfect condition for scientific examination. Zema received his full reward.

'There can be no doubt that death was brought about by decompression combined with rise in temperature,' Millot wrote in his account of the event, published in *Nature*. '. . . It must also be noted that the *Latimeria*, which appeared greatly distressed on its arrival at the surface, seemed to have recovered appreciably after an hour or so and passed the rest of the night quite comfortably without any too obvious inconvenience. It was daybreak, either the appearance of sunlight or the gradual warming of the water, which initiated the progressive discomfort which led to its death.' Millot proposed that next time the experiment was attempted, the coelacanth should be incarcerated at a depth of 150–200 metres, and only hauled up for observation or to be photographed.

Unfortunately, there wasn't a next time. During Millot's reign over the Comoran coelacanths, repeated efforts to capture or film a live coelacanth failed, and Zema ben Said was the only Comoran fisherman to claim the £200 award.

Four years after his dramatic flight to the Comoros, J. L. B. Smith had been out of the coelacanth picture for some time. While his examination of the first coelacanth had been remarkably thorough as far as it went, by virtue of its missing soft parts, he had been unable to make much of a contribution to knowledge of its internal workings. After the French got a specimen of their own, Smith left the close anatomical work to Millot and his team, and *Malania* was put on display at the Department of Ichthyology.

He never, however, ceased to feel he had some sort of dominion over the creature that he had, after all, introduced to the world. He launched a vehement attack on what he saw as French profligacy with regard to the coelacanth. He used that most august of arenas, the letters page of *The Times* of London, on 4 June 1956:

Sir,

The tragic presumed loss of Captain Eric Hunt in shipwreck not far from where he achieved fame in 1952 by reporting the second coelacanth at the French Comoro Islands brings forcibly to mind that it is now well over three years since this unique ichthyological survivor was shown to live there. Owing to circumstances it was at that time necessary to seek complete specimens for scientific purposes, and it will be remembered that not long after this discovery the French prohibited foreign scientists from searching for coelacanths in their waters, proposing instead that this should be done by an international expedition under French direction ...

The Comoros are of unique structure, with steep slopes that plunge abruptly to great depths in the ocean ... All the evidence indeed indicates that there cannot be any high population of coelacanths there. If, as is quite likely, these are the only coelacanths in existence, there may well be only a few hundred of them in all. In the past three years the French have got ten specimens, more than enough for full scientific examination. Science and the world are no longer crying for dead coelacanths ...

With my fundamental interest in this remarkable creature I

have become increasingly uneasy about the situation affecting coela-
canths . . . If a herd of dinosaurs were discovered in some remote
jungle the world would rightly recoil in horror from a policy of
rewarding natives to slaughter as many as possible. The situation of
the Comoran coelacanths is in reality no better, and the present pol-
icy is debasing a once important scientific quest to the level of sense-
less slaughter of one of our most precious heritages in biology.

The French own the Comoros, but the coelacanth belongs to
science and to mankind. French scientific authorities carry the
gravest responsibility in this matter. The policy of rewarding natives
for catching coelacanths should immediately be reversed, or modi-
fied, to that of a severe punishment for killing one . . .

I am, Sir, yours, &c.,

J. L. B. Smith

The response was rapid and indignant. Gavin De Beer,
Director of the British Natural History Museum, was the first
to weigh in. The following day, his response was published in
The Times:

Professor J. L. B. Smith's letter today implies criticism of French
scientific methods which is not deserved. When he says that ten
specimens are more than enough for scientific investigation he
appears to forget that only two of these are female, neither of which
is in a condition to show whether the fish lay eggs or give birth to
live young . . .

. . . When Professor Smith states that science and the world are
no longer crying for dead coelacanths he forgets that, while he may
possess a specimen, other museums such as this have need for one,
and have been promised specimens by our generous French col-
leagues.

The next week, Millot, the main target of Smith's attack, wrote in his own defence to *The Times*, managing, at the same time, to get in a few, possibly deserved, below-the-belt blows at the South African:

> *If Professor Smith were an anatomist he would realise that, for an extensive study, anatomical, histological and chemical, of the kind which we are pursuing on coelacanths of both sexes and all stages, a dozen specimens (which we do not even yet possess) are hardly sufficient ... I am in a position to give your readers a full guarantee that not one coelacanth caught has been superfluous ...*

He paid tribute to the co-operation of the Comoran authorities, before directing one last jab in Smith's direction:

> *In particular they [the Comorans] must be credited with having protected the coelacanths from being dynamited with depth-charges, which, though hardly believable but nevertheless true, Professor Smith twice proposed to do.*

By 1960, Millot had clearly decided that he had enough dead coelacanths of his own, and started to give spare ones away. The British Natural History Museum was rewarded for its loyalty in *The Times* battle against Smith with Specimen Fourteen; Specimen Twenty-one went to the Zoological Museum in Copenhagen. All of these were presented on the understanding that they were for exhibition purposes only, and not to be dissected for research.

After a few more years, Millot apparently tired even of his own generosity. In 1962, Dr Georges Garrouste, who had been involved in the battle to keep Zema ben Said's fish alive, wrote to J. L. B. Smith, complaining that, three months after the event, the *Institut de Réchèrche Scientifique* at Tananarive had yet to acknowledge receipt of a coelacanth he had sent them. 'In spite of our requests [they] have not yet rewarded the fisherman. In short, as a result of this negative attitude and apparent lack of interest, it has occurred to me that other scientists might be happy to benefit from such a capture.' He went on to offer Smith Specimen Twenty-six, a massive female, 1.8 metres long (five foot eleven inches), weighing a hefty ninety-five kilograms (209 pounds). JLB, however, declined, and suggested the fish be offered to the American Museum of Natural History, which gratefully accepted. There was little more he could add to the detailed work already being undertaken by the French, he felt. According to the ever loyal Margaret Smith, 'he always maintained that the first coelacanth had given him more than any man could hope for in a lifetime.'

The coelacanth had occupied the first place in Smith's thoughts for nearly two decades. With his refusal of a fresh specimen, came an acceptance that it was no longer his fish, but rather '*Notre Coelacanthe*'. Typically, however, he did not slow down. *Old Fourlegs – The Story of the Coelacanth* was published in 1956. William Smith maintains that JLB went out fishing one day in Knysna, then returned home with the whole book in his head and proceeded to write it in ten days flat. It was published in London and New York (under the title *The Search Beneath the Sea*) and translated into German, French, Russian, Estonian, Afrikaans, Slovak and Dutch. JLB dedicated the book to 'Miss M. Courtenay-Latimer, one of South Africa's most able women'. Its opening sentence is characteristic of Smith, a scientist to his core: 'These are wonderful times, and it is thrilling to be living now, though it would thrill me even more to know that I could still be here a hundred or a thousand years hence, for this immediate future promises to be of intense interest, even excitement, certainly to the scientist.'

The Smiths continued working on new editions of the *Sea Fishes* book and, after an expedition to the Seychelles, this time accompanied by William, they published *The Fishes of the Seychelles*. After 1957, JLB no longer undertook overseas expeditions, although he continued to travel to conferences abroad until 1960, when, out of consideration for his always fragile health, he stopped travelling altogether and dedicated himself to his work, and to terrorising his staff and pupils.

JLB undoubtedly had a softer side, but as he got older it was allowed to show less frequently and his reputation for intolerance blossomed. Shirley Bell, at that time a young writer working for an angling magazine, was among the honoured few that he always had time for, and the JLB she remembers is a very different man from Peter Barnett's stern professor. She received a letter from the Smiths, praising one of her articles. 'That began our friendship,' she remembered. 'He wrote me long letters in his neat spidery handwriting ... discussions about his articles that I was publishing in the magazine, news of their current activities in ichthyology, bits of advice, incisive comments in reply to my questions, gentle concern ... wonderful letters.

'Nothing was too much trouble for him,' she explained. 'He had an instinct for thoroughness and he noticed the smallest thing that he considered had been well done. He had a reputation for remoteness and even for eccentricity, and he certainly didn't suffer fools gladly, but that was not the side of him I saw in the few years of our friendship.'

He also remained close to Marjorie Courtenay-Latimer: his archives are full of typed carbon copies of warm letters to her, mainly, inevitably, about fish. His other great companion and fellow fish lover, a fox terrier called Marlin, was born in 1959. They were seldom apart. Marlin accompanied JLB wherever he went; in the car, in the boat, on long walks around Grahamstown.* Marlin even accompanied the Smiths on a trip to Marjorie Courtenay-Latimer's Bird Island in 1964, a trip he clearly enjoyed, as JLB explained in a thank-you letter to the lighthouse keeper and his wife soon after their return. 'We have been to many unusual places in the course of our work, but I do not think we shall ever forget Bird Island. We had a not unpleasant trip back to the mainland, Marlin, I think, suffered most, he hated the motion of the boat. This was probably due to the fact that he had thoroughly exhausted himself chasing rabbits, indeed his feet were sore for days afterwards.'

In the same year, and probably before Marlin's feet had even had a chance to recover, another strange chapter in the coelacanth story opened. Just as with the mysterious appearance of the Tampa scale in 1949, these new discoveries indicated that the coelacanth might indeed exist in far-flung corners of the world. An Argentinian chemist called Dr Ladislao Reti visited a small village church near Bilbao, on the Atlantic coast of

---

*In 1993, on the very route JLB and Marlin used to tramp daily, scientists from Rhodes University unearthed the fossil remains of seven coelacanths, among a multitude of other fish fossils, on the site of a prehistoric lagoon, only two miles from Grahamstown.

Spain. There he saw a strange silver model of a fish hanging on the wall as an *ex voto* – a prayer of thanks. It was about four inches long, and beautifully crafted, with the coelacanth's unmistakable paired, lobed fins, extra dorsal fin, and 'puppy-dog' tail, as well as an inter-cranial joint of its own, which unclipped to reveal the hollow centre of the ornament. Dr Reti bought the model and took it home, where he gave it to an ichthyologist, Dr Pablo Bardin, who identified it positively as a coelacanth.

Shortly afterwards, another – even more exquisite – silver coelacanth was bought privately by Maurice Steinert in Toledo, a short drive from Bilbao. Both sculptures were taken to various silver experts, all of whom were convinced that they were of Meso-American origin, probably dating back to the seventeenth or eighteenth centuries – certainly long before world attention was focused on the coelacanth. It seemed highly unlikely that the craftsmen could have crossed the world to the Comoros for inspiration: had their model, perhaps, lived a little closer to home?

JLB devoted the last years of his life to the fish books and to the Department of Ichthyology. His determination never flagged: from the time he first became interested in ichthyology, he published over 500 papers on fish, and named 370 new species. Four honoured his wife and partner, Margaret: in dedicating *Pseudocheilinus margaretae* to her, perhaps untypically he allowed his private emotions to show: 'This exceptionally beautiful creature is named as a small tribute to my wife,' he wrote, 'whose contribution to all phases of our work is probably greater than my own.'

Honours were showered on J. L. B. Smith: he was made a

Silver *ex voto* of a coelacanth, found in Spain, and believed to be of Meso-American origin.

Fellow of the Royal Society of South Africa, an Honorary Foreign Member of the American Society of Ichthyologists and Herpetologists, and a Corresponding Foreign Member of the Zoological Society, London, among others. But latterly, he refused some of the honours offered to him: 'Give it to a younger man who is still climbing and who would appreciate it – it would be wasted on me,' he said in response to one. He agreed, however, to receive the degree of Doctor of Science *honoris causa* from Rhodes University in April 1968.

For the last two years of his life, JLB felt his mental powers were deteriorating. Margaret was taking over an increasing burden of her husband's work at the Department of Ichthyology. His eyesight was failing and he was terrified of having a stroke. One of his greatest university friends, the former Minister of Internal Affairs, Donges, had died after a stroke a couple of years earlier. Smith, always conscious of his ailments, dreaded becoming bedridden and a 'useless hulk'.

At the end of November 1967, he handed his secretary, Jean Pote, a large cheque. It was twice as large as the Christmas bonus she had expected. 'I want you to keep it,' he told her. At Christmas, he kissed his daughter-in-law, Gerd, Bob's wife, on the forehead. 'I was surprised. It was the first time he had shown any sign of affection,' she recalls. He was similarly demonstrative when he saw his old friend and coelacanth compatriot, Marjorie Courtenay-Latimer late that year. 'Whenever he came to East London, he would have lunch with me. So this time, we had eaten, and when I went out to the car to say goodbye, he threw his arms around my neck and kissed me on the cheek. I thought it was very strange – he had never done that before. Then he said, "Well, lass, you're doing a fine job, carry on." I laughed and thought what a funny thing it was for him to say.'

On 8 January 1968, aged seventy, at his home in Grahamstown, J. L. B. Smith took a fatal dose of cyanide. It was carefully planned and neatly carried out. He left two notes. The first was addressed to Margaret:

> *Goodbye my love, and thank you for a wonderful thirty years. I am going upstairs to the servant's room. Careful. Cyanide.*

In the second, he explained himself. He had typed:

> *. . . For some years I have suffered from severe mental depression . . . the sight of one eye has almost gone . . . back pressure is proving troublesome . . . I live in perpetual fear of becoming bedridden*

J. L. B. Smith with Marlin at Knysna, shortly before his death in 1968.

and helpless . . . I prefer to take this way out, probably only a brief anticipation of nature.*

J. L. B. Smith was mourned across the world, by scientists, anglers and friends. Margaret received a flood of tributes: 'I have always treasured our friendship and I have especially appreciated his kind attention and very wise advice he so readily gave me,' wrote the scientist, R. Liversidge. 'I owe him a lot and far more than most realised . . . for the philosophical attitude he held towards science, administrators and his fellow men.'

* Both of these notes have disappeared from the Smith files.

Humphrey Greenwood of the British Museum described how, 'as a student I stood in awe at the work he did and had done. Now, having had many years' experience of similar research, the awe is further emphasised.' At the opening of an exhibition at the East London Museum to celebrate J. L. B. Smith's life and work, the president of the CSIR, Dr H. J. van Eck, compared the Smiths' scientific collaboration to the Curies' and the Webbs'.

'I see him now as I always knew him, lean, virile, fearless and absolutely honest in all his dealings with both men and with scientific facts and theories,' lamented a friend, Dinnie Nell. 'I would have turned to him before anybody else amongst my friends and acquaintances and been quite sure of warm sympathy, advice and any needful assistance,' the Ichthyology Department librarian, Doris Cave, wrote.

His old friend and coelacanth companion, Marjorie Courtenay-Latimer, was overwrought. 'How honoured we were to know him,' she wrote to Margaret Smith. 'He was a brilliantly clever man, and I was very, very lucky to have known him.'

VIII

IN COMORIAM

By 1963, the French had decided they had enough specimens to work with. They moved their base of operations to a special laboratory at the Muséum National d'Histoire Naturelle in Paris, and handed over the distribution of further fish to the Comoran authorities. The Comorans were delighted, and instituted a practice whereby the fishermen had to sell coelacanths to the government for a fixed reward of £100, and it would, in turn, dispose of them how they wished. Picture pamphlets were distributed to all the schools in the Comoros, explaining what the coelacanth was, and urging the children to go home and tell their parents. The sole Air Comoros plane was put at the disposal of the fish: whenever one was caught, the plane would be dispatched immediately to bring it back to the government freezer in Moroni. The French still monitored the catches, and for the next decade, until the Comoros declared their independence and the French left, records show an increased catch. Nine coelacanths were fished in 1965, and six the following year.

*

It is a testament to the remoteness of the Comoros that the coelacanth was not known to live there before 1952. Legend has it that King Solomon married the Queen of Sheba on the remote Indian Ocean islands a thousand years before the birth of Christ, but evidence of a more archaeological nature dates the arrival of the first settlers to be during the first century AD. They sailed from Indonesia, 6,000 miles across the Indian Ocean in narrow dugout canoes, several times longer than the *galawa* used by fishermen today, but similar in shape, with the same double outriggers. They were men of the sea, and when they colonised the islands of the western Indian Ocean – first Madagascar and then the Comoros – they brought with them their skills and techniques.

They found four perfect jewels, strung across the neck of the Mozambique Channel. The volcanic soil was fertile, and the sea full of riches. It is not known when they first caught the strange fish they called *gombessa*, but it is possible they have been fishing it up for hundreds, perhaps even thousands of years. Some of the time, they just threw it away. It was *nessa*, the oil fish, they wanted; *nessa* that commanded a good price at the market, and still does to this day. It is bought for its medicinal qualities – its oil-rich flesh acts as both a laxative and a mosquito repellent. *Gombessa*, on the other hand – which in Swahili translates as something taboo, strictly forbidden – does not taste good, and its oily flesh is a violent purgative: its value lies in its ancient pedigree, which the Comorans neither knew nor cared about.

Over the centuries, the first Comorans were joined by Arabs, Shirazis and African immigrants: today, you can clearly see the mixed ancestry in the Comorans you meet. Soon after the birth of Mohammed in AD 570, a Comoran emissary returned from Arabia with a caliph, who converted the country to Islam.

By the fifteenth century, the islands were a centre of maritime trade, ruled over by power-hungry sultans, constantly at war with one another. Ships plying the wealthy East African coast stopped off in the Comoros to trade rice, ambergris and slaves. Inevitably, they attracted the attention of pirates, who used to hide in secret coves, ready to pounce on the overladen trading vessels. By the beginning of the last century, however, the Comoros had subsided into civil war, as the sultans battled with one another to supply slave labour to eager French plantation owners on Île de France (now Mauritius) and the Seychelles.

The end of the slave trade and the opening of the Suez Canal in 1869, effectively banished the Comoros to a back seat in the theatre of international trade. By the close of the century, only two or three ships stopped at the islands each year, and the country slipped into penury. France stepped in, and in 1946 the Comoros became a formal colony. French loaves and *pains au chocolat* were integrated into the Comoran diet of fish, rice and cassava, and even today, on every street corner in the capital, Moroni, you can find a Comoran lady sitting in front of her bag of crusty baguettes.

They were among the smallest and most remote of French colonies: even the French foreign office probably had little idea of their existence, before attention was sharply focused on them after the diplomatic débâcle of Smith's appropriation of the 1952 coelacanth. Les Comores were nothing but a small drip, drip from Quai d'Orsay coffers. When the discovery of *Malania* was announced to the world, it was not only JLB who had to reach for an atlas.

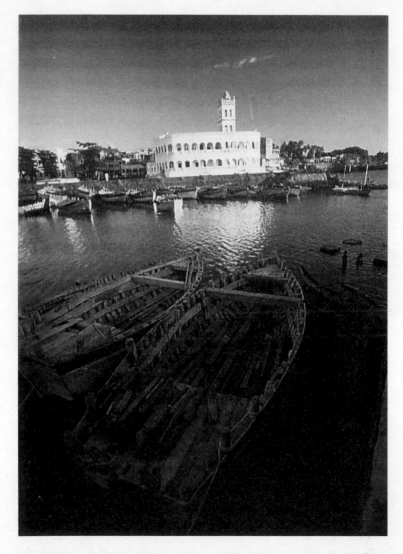

The Friday Mosque in Moroni, Comoros.

The coelacanth provided a boost to both the Comoran pride, and the economy. Sidi Bacari Papa, the taxidermist on Anjouan, was kept busy, as favoured visiting heads of state and other dignitaries were presented with a stuffed or frozen coelacanth: the United Nations received one, as did President Mitterrand and South African Foreign Affairs Minister, 'Pik' Botha. It was a gift no other nation had the power to bestow. The coelacanth was proudly minted on to Comoran coins and notes, it was printed on colourful pareos and T-shirts, and fashioned into delicate gold brooches and necklaces, bridal *grand mariage* presents. In December 1986, the Minister of the Interior declared it to be a 'heritage of humankind', and the Comoran nation, its warden.

The fishermen who caught a coelacanth benefited too. The islands are extremely poor, and for most people, subsistence farming and fishing is a way of life. Few people starve, but fewer still have money to burn. The society is strictly stratified: at the top of the pile sit the Sharifs, the descendants of the Prophet Mohammed. Below them are the intellectuals, the *notables* of the mosque, the civil servants and teachers. Then come the farmers and, below them, on the very lowest rung of Comoran society, perch the fishermen. They are regarded as coarse – loud voiced and fond of a brawl. On the whole, the sons and grandsons of fishermen themselves become fishermen, and marry the daughters of fishermen. Even if they work hard to become intellectuals, they are still marginalised by society. There is a Comoran saying: 'Even if you make fish soup, it still smells of fish.'

For a while, the coelacanth changed all this. Suddenly, the fisherman who caught one was a local hero, fêted by the *mzungus*, white men who hovered around the fishing villages,

The coelacanth as a Comoran cultural symbol: on coins and banknotes.

urging them to hook a *gombessa*, and eager to find out all the details when they did. Apart from Captain Goosen's chance trawl in 1938, every coelacanth had been caught by Comoran fishermen using traditional techniques.

The fishermen of Itsoundzou, a small village on the south-western coast of Grande Comore, are on the black, lava rock beach in time to launch their small *galawa* in the last, pink light of day. They paddle quickly out to sea, small wiry men in tatty T-shirts and holey shorts, battered palm-frond hats on their heads. A short way from the shore, they stop briefly to let out a line, and within a minute they bring it in again, with a small silvery fish, a *roudi*, on the hook, which they use as bait.

The fish is laid carefully in the bottom of the *galawa*, while the fishermen squat, their knees under their armpits, on the two benches which cross the narrow boat. They row out to sea with strong strokes, wielding their single short paddle first to one side, then to the other. Their movements are swift, but controlled – even with the spidery outriggers, the *galawa* is surprisingly unstable, and a sudden movement to one side could easily topple it.

As the sun slinks beneath the horizon, vertical clouds rise out of the sea, low-lit like a ghostly Gotham city. It is soon dark; there are no lights on the shore, save for the occasional car wending its way along the coast road (even today there is no electricity this far south on Grande Comore), yet the fishermen know exactly where to go. After a lifetime of nights on this inky ocean, they are intimately acquainted with every reef and cave, every dip and rise of the ocean bed. About 500 metres from shore, the fishermen carefully lay their paddles across the *galawa*, and prepare the line. They are searching for *nessa* (*Ruvettus pretiosus*, the oil fish), a large and ugly fish which lives in the semi-deep, and patrols the same hunting ground as *gombessa*, the coelacanth. The fish are similar in size and weight, and are caught in the same way.

The moon is a slim crescent, and casts a muted glow on the black water; *nessa* and *gombessa* don't venture out at full moon, when the sky is too bright. Quickly, the fishermen tie two flat black stones, collected from the beach, about eighteen inches (forty-five centimetres) above the baited hook, a method they call *mazé*. They then let out the single line until it touches the seabed, measuring it as it goes; each arm's length is a metre. When the stone sinkers hit the bottom, they raise the line a metre or two, before jerking free the rocks. They then jiggle the bait along the ocean bed with a quick see-sawing action of their arms, feeling its movement through their fingers.

There is a wonderful sense of calm. The only sound is the occasional muted ripple as the fishermen manoeuvre the paddle tucked under their arm, and a swift burst of action when a fish is caught. The *galawa* are huddled together, but the men seldom talk to one another, and there is a comforting sense of community in the silence. It is a dangerous job – on nights when the wind is strong and the waves high, fishermen have been known to disappear, never to return. Through each long night, they stay on their boat beneath the star-filled southern skies, until the early hours, when they paddle back to shore, pull their *galawa* over the smooth, rounded black lava rocks, and carry their catch up to their huts to wait until the market opens at daybreak. If they have hooked a *gombessa*, however, there is no time to waste – they have to find one of those crazy *mzungus* before the fish starts to decay in the morning sun.

The fisherman who sold a coelacanth suddenly had more money than most people earned in five years. Even today in the Comoros, £100 is a lot of money, a jump start on the route to a *grand mariage*, the cornerstone of Comoran society and the method by which an ordinary man of the village becomes a respected *notable*.

Mzé Lamali Hila started fishing when he was ten years old. He claims to be over a hundred now, and still climbs from his sagging rattan bed into his *galawa* for nightly fishing trips. These days he doesn't catch so much. 'I have caught four *gombessa*,' he says with pride. 'And I remember when I was a small boy my father caught one. It doesn't taste good, but we used it as medicine for many diseases: upset stomachs, spots, hip problems. The first one I caught was a long time ago,' he recalls. A small crowd has gathered at the entrance to Mzé's ragged, palm-weave hut, and the old fisherman starts playing to his audience, telling his story in a melodic, swooping voice. 'I was doing *mazé*, I had dropped my stones on the bottom, and then I felt a bite. I started pulling it up to my boat, very slowly; it took almost two hours and it was very heavy. When I got it to the surface, I put a very big hook through its mouth to attach it to the boat. It made a wave as I pulled it to shore. In those days, before the whites came, it had no value – you couldn't sell it at the market, so the next day, we cooked it. It was not delicious, the flesh is too fatty, not good, and many people didn't eat it – only some children and villagers from up the hill. If you eat *gombessa* or *nessa*, you have to have a shower,' he recounts to the amusement of his audience. 'It gives you diarrhoea.' After 1952, he fared much better. He sold his second coelacanth for 100,000 CFA (almost £200); his third and fourth for 50,000 CFA and 40,000 CFA respectively. With rewards like that, every fisherman was eager to hook a *gombessa*. While there is no evidence that they can fish them on demand, the monetary incentive enticed more fishermen to go out at night, leading inevitably to a bigger catch.

The Comoran government was doing a roaring trade in coelacanths. In 1966, biologist Keith Thomson was working at the Peabody Museum of Natural History at Yale University, when they received a circular letter from the Comoran government, offering any zoological institution the opportunity to purchase their own coelacanth for the bargain basement price of $400, plus shipping. The demand, inevitably, was high, and today a coelacanth occupies a place of honour in most of the world's premier natural history

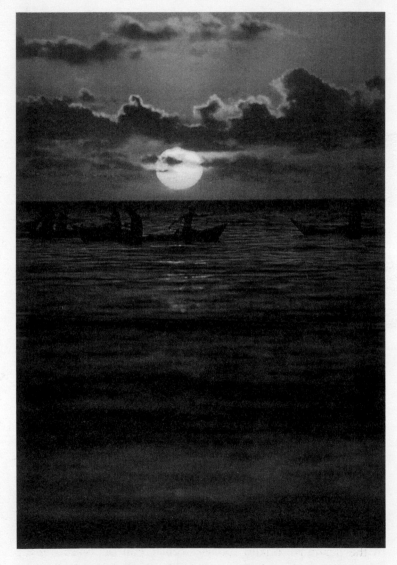

Comoran fishermen at sunset.

museums. The Peabody managed to get hold of the first fresh-frozen coelacanth, which was displayed, prior to dissection, in an ice-cream freezer case. 'Most of Connecticut filed past,' Thomson reported. 'It was a miniature version of what J. L. B. Smith must have endured. The Great Hall of Yale's Peabody Museum came to resemble nothing less than Lenin's Tomb.'

Once the French had left for good in 1972, permission was granted for overseas expeditions to hunt for their own coelacanths around the islands. What they all wanted was a live specimen: apart from a few dying fish paddling around in the shallows, no one had seen a living coelacanth.

The potential rewards for the first images of a live coelacanth were so great, that photographers went to dramatic lengths to obtain them. Only a month after the capture of '*Notre Coelacanthe*' in 1953, an Italian zoological expedition, led by Franco Prosperi, claimed to have photographed a live coelacanth off Mayotte. In his book, *Vanished Continent*, Prosperi tells of how he leaned over the side of the dinghy and peered into the shallow water. 'For some minutes a strange fish had been occupying my attention. It was resting on a madreporic [hard coral] rock about forty feet below the surface. As I say, it was not swimming, but was lying on the corals, too lazy even to move a fin ... I carefully studied its squat body, its elevated trunk, and its uniformly dark colouring ...' He recognised it by its strange fins to be a coelacanth, and with his 'heart pounding', dived into the water and swam towards the putative coelacanth. 'I saw the outline in the viewfinder,' he wrote. 'I noted the sturdy appearance of the body; the fins protruding like small spatules from the end of the fleshy peduncles ...' He took a picture, but sadly for him, 'The sudden click and the subsequent whirring of the mechanism galvanised the fish into life. Turning with a speed incredible in such a ponderous creature, it headed rapidly for the sea bottom. While I was trying to follow it I saw Fabrizio swim past on my right, his body rigid behind his gun.'

Fortunately for the coelacanth, Fabrizio didn't manage to get a clear shot. Soon afterwards, the French issued their edict forbidding foreign scientists from catching a coelacanth, and the coelacanth population was spared the Italians and their underwater armoury. The single photograph of the coela-canth was printed in *Vanished Continent*, with extravagant claims of it having been the first 'living' picture. Scientific fraud antennae immediately started to twitch: the picture is unclear, and there are various other clues indicating that it was a fake encounter. For a start, no other coelacanth has ever been seen off Mayotte, or in such shallow water; a live coela-canth is not a 'uniformly dark colour', but covered in white splodges; and it has never been observed resting, or even touching the surface of the coral. Most coelacanth experts believe the picture was somehow falsified.

Thirteen years later, a French photojournalist called Jacques Stevens, sold a story to *Life* magazine, in which he purported to have filmed a coelacanth swimming in the 'dim and aerie' depths of the ocean. He was diving at night when 'out of the spooky darkness at about 130 feet below the surface, here came a coelacanth'. The fish was 'mucus-covered', and its 'huge phosphorescent eyes glared at me'. Unfortunately for Stevens – and posterity – his movie camera jammed, and the flash of his stills camera appeared to frighten the coelacanth, which, he claimed, swam off after he had only taken two pictures.

When the story appeared, scientists again quickly recognised that all was not exactly as it was intended to appear. The fish was photographed against coral that is only found in shallow water, and the pictures themselves were well lit and clearly taken in sunlight. In addition, they saw, on the coelacanth's snout, marks consistent with the effect produced by the rubbing of a fishing line; and it had cloudy eyes and exhibited clear signs of stress. It was apparent to the scientists that this denizen of the deep had, in fact, been caught by a fisherman in the usual manner, then taken to the shallows, where Stevens had photographed it in its dying stages.

Another Frenchman, local dive master Jean-Louis Geraud, made no bones about the coelacanth he filmed being on its last legs. 'It was wonderful,' he said, in heavily accented English, 'It was exactly like I see a dinosaur in my garden, but a nice dinosaur because he swim like he dance.'

In 1972, a joint Franco-Anglo-American expedition failed to catch their own fish, but were on the scene when two coelacanths were caught by local fishermen. The first, a large female, was dissected on the spot and found to contain nineteen huge eggs, the size of grapefruits, weighing around three-quarters of a pound each (300 to 350 grams). These were the largest fish eggs in the world, and further suggested that the coelacanth gave birth to eggs instead of live young.

But in 1975 – thirty-six years after J. L. B. Smith examined his first coelacanth – that theory was finally knocked on the head. The Comoros, under the leadership of Ahmed Abdallah, had declared unilateral independence from the French, and the embargo against international scientific institutions examining their specimens no longer held. The American Museum of Natural History immediately started their dissection of Specimen Twenty-six, the vast coelacanth that had originally been offered by Dr Georges Garrouste to J. L. B. Smith, but which he had declined. When they opened up the fish, they were amazed to find five almost fully developed *Latimeria* pups, close to thirty centimetres long, and each attached to the remnants of a large yolk sac.* The fossil evidence had at last been proved: *Latimeria* was a live-bearer.

This knowledge had important implications. If the early coelacanths had also born live young, in doing so, they would have predated the first mammals by 100 million years. But at the same time, it indicated a slow population growth rate: the coelacanth is believed to have a gestation period of over a year – if it produces only five young, then the rate of renewal of a small coelacanth population is necessarily slow, and any increase in predation might wipe them out completely in a short period of time. A group of scientists, led by Eugene Balon of Guelph University in Canada, went further, and surmised that the small number of embryos – versus the large number of eggs – was evidence of oophagy, or inter-uterine cannibalism, which is also common in sharks. This theory was immediately refuted by another group of coelacanth specialists – but the truth was that none of them knew for certain. Calls for the capture of a live coelacanth which, it was thought, would enable such mysteries to be studied closely, increased in volume.

The Steinhart Aquarium raised the stakes by offering the successful fisherman not only a monetary award, but a two-

---

*On 24 March 1975, the East London *Daily Dispatch* reported that Californian scientists had dined on fillets of this particular coelacanth. They had defrosted some frozen muscle, cooked and eaten it. According to Dr John McCosker, 'It was absolutely delicious.'

week, all-expenses-paid trip to Mecca. It still didn't do the trick. Coelacanths, it appeared, could not be kept alive for any length of time at the surface.

In August 1975, a young revolutionary named Ali Soilih overturned the first Comoran government of Ahmed Abdallah, and set his country off on a roller-coaster ride of coups and counter-coups. Initially, Soilih brimmed with reforming zeal, but after a series of setbacks and determined opposition from the conservative elders, he descended into madness. He sacked the civil service and turned the running of the country over to teenagers with guns.

Two years later, a BBC cameraman, Peter Scoones, was sent to the Comoros to try to get live coelacanth footage for David Attenborough's *Life on Earth* series. After several frustrating weeks of trying to film in the ocean using an ROV (Remote Operated Vehicle) at a depth of 1,000 feet, the closest he had got to a coelacanth was a stuffed model in the Presidential Palace. 'We were coming back from a dive one evening, in the Comoran Navy's only boat, and I looked at the island and it was red, as if it was sunset,' he recalled. 'Only we were on the wrong side of the island. I realised I was looking at rivers of molten lava – the volcano had erupted.' Karthala, with its huge crater, dominates the island of Grande Comore. It erupts, on average, every twelve to fifteen years, destroying villages and depositing black lava into the ocean.

'Since I hadn't got any coelacanth footage, I thought I may as well film the volcano,' Scoones continued. 'I spent the night climbing through the rainforest, dodging clods of red-hot material which kept flying through the air. I got back to the hotel (appropriately, the Coelacanth Hotel in Moroni) at dawn, and no sooner had I got into my room, than there was a knock on the door.' A coelacanth, he was told, had just been caught in a nearby village and was still alive.

'I rushed to the village to find it tethered under the shade of some pirogues. I carried it back into the water and tried to revive it by passing water over its gills. It was clearly exhausted, and one of its fins had been almost severed. It kept trying to bite my hand, which was actually quite useful, because then I could manoeuvre it into a good position to film. The only problem was that it seemed to want to stand on its head or swim upside down. Still, I got some good pictures.'

Scoones's adventure was not over. A few days later, he was in Moroni where he noticed young militiamen patrolling the streets. He walked past a two-storey government building. Paper was floating from the windows to the ground, where he saw it being burnt on a small bonfire. 'I didn't really know what was going on,' he says. 'I asked some chap, who said it was a coup, but it didn't really look like much. Like everything else on the islands, it was pretty disorganised.' Unbeknownst to Scoones, he had witnessed Ali Soilih's last attempt to brand his presidency – by burning the government archives. With the treaties, laws and records, went all the papers relating to coelacanth catches over the five years since the French had left. In coelacanth history, the Soilih years are a complete blank.

Ali Soilih was overthrown in 1978 by a mercenary force led by the infamous soldier of fortune, Bob Denard, who had been hired by the first President of the Comoros, Ahmed Abdallah. For the next eleven years, Denard and his men were a constant presence on the islands, overseeing their security. In time, they came to be feared and hated by the majority of the population. Coelacanth specialist Robin Stobbs tells of going to the Comoros on a Smith Institute research expedition. 'We were sitting in a fairly empty plane, when a distinguished, silver-haired man came back from first class, asked who we were, and introduced himself as Colonel Bako [Denard's pseudonym in the Comoros]. It was only later that we realised who he really was.'

The Smith Institute group was given an escort by the mercenary-led Presidential Guard. 'I couldn't work out at first why the locals were so uncooperative,' Robin Stobbs continued. 'It was only when we met them without the mercenaries, that they opened up.'

Adventure and disorder were also on the menu for the Explorers Club, led by the indomitable coele-enthusiast, Jerome Hamlin. A Yale philosophy graduate, and inventor of the first domestic robots, he had become intrigued by the coelacanth in 1984, when he joined a New York Aquarium beluga whale hunting expedition, as a photographer. 'The procedure was quite fascinating, and at the conclusion I asked various aquarium experts and observers what the most challenging and scientifically valuable collecting project would be. They answered, "the coelacanth".'

It was as good as a lure for Hamlin. In 1986, he went to the Comoros for the first time, to sniff out the possibility of capturing a live coelacanth for the New York Aquarium. Two days after he arrived, he was having tea at the Coelacanth Hotel, when a waiter announced that a coelacanth had been caught. 'We were down the coast road in a blur, our hearts pounding,' he recounts. 'The fish was dead, but very fresh and its eyes still glowed. I have never seen such a beautiful creature. We bought it from the fisherman and persuaded the chef at the hotel to store it in his freezer. I was careful from then on not to order "fish of the day"!'

A few days later, he was woken by a knock on his bungalow door and an excited whisper: '*Un coelacanthe vivant.*' He was taken to the shore, where a fisherman was waiting with a living coelacanth tied to his boat. Hamlin climbed into the canoe and tried to keep the fish from bumping against the side. He then floated a folded rubbish bag over the fish's eyes to shield it from the rising sun. 'As soon as I saw it, all thoughts of photographing the creature flew from my mind: all I wanted to do was to keep it alive,' Hamlin recalls. He summoned the French diver, Jean-Louis Gerard, who swam down with the fish, and secured it to the ocean floor with a rope threaded through its lower jaw. Sadly, the coelacanth was unable to survive the adventure, and died overnight. It was taken to the US Embassy freezer where it stayed, next to the body of a small American child who had died of malaria, until the next available flight, when it was transported to America in a coffin lined with pallets of dry ice flown over especially from France.

After two such near misses, Jerry Hamlin was hooked. He returned to the Comoros the following year, with crates of equipment, including an enormous fish transporter tank, refrigerator units, oxygen tanks, portable generators, and a couple of float bags. At that stage, he still believed it was important to get a live fish in captivity, to learn how it lived and how to husband them. With a team from the New York Aquarium, he set out day and night to find one, with little success. 'But there was no way I was going to give up,' said Hamlin. 'I was determined to keep coming back until I got a fish, and I started to put in place the structures for a more permanent coelacanth-monitoring team. One afternoon in 1987 I noticed a large Comoran man – with the build of a silver-back mountain gorilla – hanging about the hotel pool area. We got chatting, and it transpired that he worked there as a security guard several days a week and went by the name of Mombassa, after the town where he had been a boxer several years before. He had also acted in two John Wayne movies – he was the chief safari boy in *Hatari*.' It transpired that Mombassa was a famous character on the Comoros, and had worked with a Japanese coelacanth-hunting expedition the previous year. Hamlin hired him as his local Mr Fix It, and Mombassa became his partner in adventure for the next ten years.

Hamlin returned to the Comoros in 1988, to try again. This time he operated without the New York Aquarium. A series of misunderstandings had resulted in some major rows in the coelacanth world, which had prompted the aquarium to pull out of the project, leaving Hamlin more or less to his own (often extraordinary) devices.* He had become consumed with the idea that he had to save the coelacanth, and to that end, he changed his aim from capture to conservation. 'The idea was to resubmerge the fish in steel cages to their capture depths, in the hope of resuscitating them,' he explained. On five occasions specimens were returned to depths of 90 to 150 feet (30 to 50 metres). One fish managed to survive for five days, during which time it was fed chicken by divers, before dying, possibly as a result of an eel attack. Another bit John-Louis Geraud on his hand; a third was lost, after Mombassa

---

*A colleague of Hamlin's had placed an advertisement in an angling magazine, appealing for volunteers to monitor the coelacanth project in the Comoros. Various coelacanth scientists read this as an indication of the New York Aquarium's desire to fish actively for coelacanths. After an exchange of angry letters in the *New York Times* the aquarium dropped out. Hamlin later made his peace with the scientists.

fell asleep on the beach and failed to notice that its marker buoy had vanished.*

In 1989, Hamlin arrived in the Comoros the day after a coelacanth had been caught and resubmerged. He went straight from the airport to the village, where he found the fish already dead, with big yellow bruises on its lips, perhaps a result of banging its head against the cage. These obvious signs of stress prompted Hamlin to drop the cage idea, and go into 'passive mode'. He installed a 700-gallon, cooled-down aquarium in a tent opposite Itsandra Beach, looked after by Mombassa. The first two fish that were intercepted were rushed to the tank upon capture, but both died before they arrived. A coelacanth caught in 1995 lived for ten hours in the facility, having survived for fifteen hours previously on the ocean surface. There were no further attempts and in 1997 Mombassa was killed in a traffic accident, temporarily halting the possessed Hamlin's efforts.

Foreign expeditions continued, with the clear aim of trapping a live specimen. An American aquarium offered to pay $40,000 for a living coelacanth, while the Toba Aquarium of Japan launched the most ambitious – and certainly the most expensive – attempt yet to capture two live specimens in 1989. The multi-million-dollar project, sponsored by Mitsubishi, used a research vessel equipped with an ROV, specially designed traps, and fishermen from Japan and the Philippines, but they still failed in their objective.

It was about that time that a worrying rumour started to circulate. It was said that the coelacanth had entered Chinese mythology, and venal Chinese herbalists were promising that

*This fish was rumoured to have been stolen by the mercenaries.

a drop of notochord fluid would guarantee immortality. Coelacanths were apparently being picked up on the black market by Japanese middlemen and sold on to Chinese practitioners who, the rumour went, were selling a single drop of the fluid for the incredible sum of $1,000. Since a coelacanth contains something in the region of three litres of clear, amber-coloured oil, this would make it an extraordinarily valuable fish. Suddenly, it dawned on the scientific community that the coelacanth was in danger of extinction. Although no one could calculate the exact size of the coelacanth population, it appeared to be shrinking. What J. L. B. Smith had prophesied in his letter to *The Times* in 1956 was coming true: the 'seemingly uncontrolled massacre' of coelacanths in the Comoros threatened to wipe out the species altogether. And their blood would be on the hands of the scientists. The emphasis, they realised, would have to change from collection to protection.

And it was clear that, in this, they were supported by the public, who regarded the continued survival of the coelacanth as important. A German magazine article had asked readers, 'Why is it worthwhile living this week?', to which a schoolchild had responded, '. . . that coelacanths still exist.'

IX

DER QUASTENFLOSSER

In 1975, a young German scientist visited the Comoros to dive for the coelacanth. After several failed attempts, he turned to his wife, and said, 'Next time, I will come here with a submarine!'

He was not entirely joking. Since reading *Old Fourlegs* as a teenager in East Germany, Hans Fricke had been determined to see a live coelacanth. From an early age, he had been entranced by fish and marine life, and loved diving which, unfortunately, he had little opportunity to pursue behind the Iron Curtain. 'More than anything, I wanted to see a coral reef,' he explains. 'I knew I would never be able to, while I lived in East Germany, so I had to escape.' In 1960, a year after he had read Smith's book, Fricke – then a young Navy cadet – left his family and boarded a train to West Berlin. The Berlin Wall was on its way up, and he knew that if he was caught by the police who patrolled the train, checking everyone's papers, the consequences would be serious. Searching for a hiding place, he ducked into a cabin reserved for women and children – where he saw a high-ranking Army officer with his two young daughters. 'I sat beside him and started talking. When the policemen came to our carriage, they recognised the officer's seniority and didn't stop. They must have thought I was his son.'

Fricke made it to the West. He went back to college, as his Eastern qualifications were not accepted in West Germany. At night, he sold newspapers in the city's night-clubs, bars and brothels. 'The women were all very charming,' he recalls with a smile. 'They were good to me because, like them, I had a street profession. I remember once, at about five in the morning, I was terribly tired but I still had fifty papers to sell. I went to a striptease place, and one of the dancers took pity on me and bought all my papers – without losing the beat.'

With the money he earned from his night job, Hans Fricke went to the Red Sea in 1961 and saw his first coral reef. He was hooked. The next year, he returned – by bicycle, across Europe to Greece, by ferry to Alexandria, and by bicycle again to the Red Sea. He cycled a total of 10,000 miles, picked up a debilitating stomach bug, and lost forty-four pounds in the process, but the struggle did little to diminish his wonder. From then on, he went back every year, to study and photograph the exotic reef life.

He never forgot, however, his desire to see a living coelacanth. In 1968, he started working with the famous Conrad Lorenz at the Max Planck Institute for Animal Behaviour in the small Bavarian hamlet of Seewiesen, and a year later, he went on a research trip to the island of Nosy Bé in Madagascar. 'There I met the French scientist, Raphael Plante, and we dived for coelacanths along the continental edge of Madagascar. Of course we had no luck.' They stopped off in the Comoros on their way back to Europe, and stayed in the Hotel

Coelacanthe in Moroni, from where they dived again for the elusive fish – again with no success. It was six years later, on Fricke's next trip to the Comoros, this time with his wife, Simone, that he decided to return with a submarine.

The love affair between marine biologists and vehicles of the deep had really begun forty-five years earlier when William Beebe, from the New York Zoological Society, became the first person to conceive of a way to explore the ocean depths. After years of examining a small area of water near Bermuda, using nets to pull up more than 115,000 animals – representing 220 species of marine life – he had become increasingly frustrated at the limitations of his method. The fish he sampled were at best the passing trade of his ocean patch – and he knew his discoveries were literally little more than a drop in the huge ocean. So he decided to invent a machine that would take him into their world, to see at first hand where and how these creatures lived.

In the early 1930s, Beebe designed the first bathysphere (from the Greek *bathys* meaning deep). It was in effect a steel ball, specially created to withstand the great pressures of deep water. Its walls were made of an inch and a half of solid steel, its windows of solid quartz six inches wide and three inches thick, and it had a hollow just large enough to accommodate two cramped people. With its spidery legs and cable fitting, it stood as tall as a man, with a small entrance hole sealed by a door that weighed 400 pounds, and an outside spotlight which could be switched on and off. Oxygen was stored in pressurised tanks, and carbon dioxide neutralised with the aid of an open tray of chemicals over which Beebe periodically waved palm fronds in order to speed up the reaction.

The sphere was lowered by cable into the often pitching ocean, while Beebe and a colleague sat inside, watching their world turn an ever deeper shade of blue. From his first dive in June 1930, Beebe was captivated by a world that he was the first living person to see. 'On earth at night in moonlight I can always imagine the yellow of sunshine, the scarlet of blossoms,' he recorded in his colourful account of his underwater expeditions, *Half Mile Down*. 'But here, when the searchlight was off, yellow and orange and red were unthinkable. The blue which filled all space admitted no thought of other colours.'

He saw all kinds of weird and wonderful creatures, many previously not known – or even thought of. There were fish and jellies of extraordinary shapes, illuminated by their own flickering lights. On one trip, he saw a huge fish with large eyes and a gaping jaw which it held open to reveal fangs, glowing with what appeared to be luminous mucus. Spots of pale blue lit its body, while two tentacles drooped from its head, one tipped by a reddish light, the other blue.

In 1934, he descended to a record depth of half a mile – many times deeper than man had ventured before, either in a dome-capped suit, or a submarine. There he saw a creature more than twenty feet long and very wide, seemingly without colour, texture, or even a distinct outline, moving silently through the dark water. He was never able to identify the strange, serpent-like beast – to this day nobody knows what he saw – and he concluded that man's knowledge of the watery deeps was indeed still very limited.

Successive generations of marine biologists latched on to Beebe's invention and began to build more bathyspheres and submersibles – as they are now known – to explore ever deeper parts of the ocean, where they encountered a huge number of new and extraordinary species. This enthusiasm filtered into the public consciousness, thanks in part to Jules Verne's *Twenty Thousand Leagues Under the Sea*, which was inspired by the early bathysphere voyages; and to Jacques Cousteau, who became famous for his films of underwater life.

Hans Fricke started looking into the possibilities of hiring a submersible soon after his return from the Comoros in 1975. But it was three years later that he began to take his quest seriously. At a conference in Geneva, he met Jacques Piccard, who had descended to a record depth of eleven kilometres (6.8 miles) in his submersible. Piccard offered to take Fricke for a dive in Lake Geneva, and as they resurfaced, Fricke turned to him and said, 'This is something I want to do in the future.'

First, Fricke looked into the possibility of bringing Piccard and his submersible to the Red Sea, but it turned out to be far too expensive. 'So I investigated whether I could purchase a submarine. But that was also too expensive,' he recounted. 'I soon realised that most of these machines were far too complicated, and needed specialists to service each of the parts. I wanted a machine which was simple enough to enable me to do my own maintenance. I started to contact people who built submersibles.'

Among the strange and interesting people he met was a Czechoslovakian engineer based in Switzerland, called Jaroslav Kahout. Fricke went to his workshop near Zurich, and was impressed by his machining techniques and by his ability to improvise, 'typical of a man from the East'.

Together, the two exiles from behind the Iron Curtain set about building a two-man submarine at the bargain basement cost of 130,000 DM (about £35,000), which the German magazine, *Geo*, had pledged to fund. 'It was very, very basic,' Fricke explained. 'I used my imagination, and Jaroslav translated that into reality. We made dummies of the interior to find out the psychological limits to space – how small we could make it without going mad.'

A yellow submarine, not unlike a beetle itself, was born and christened *Geo*, after the magazine. It had rounded, wing-like ballasts filled with pressurised air on top, and two, thick, bug-eyed portholes, one at the front and one on top – like a skewed pair of spectacles for the very poor-sighted. There was a spotlight attached outside, a base for a camera, and a long, claw-like manipulator, which could be operated from inside, and used to deposit and pick up materials from the ocean bed. There was just enough space in the cabin for two: the pilot sat behind the observer, whose eyes were on a level with the window. It was designed to be able to descend to a depth of 200 metres (660 feet). Hans Fricke was delighted with his new baby, which, for the first few years, spent most of its time in Europe and the Red Sea. 'I hadn't forgotten about the coelacanth,' he said, 'but for that project we needed a mother ship, which wasn't easy to find.'

One of his adventures to the Red Sea in the early 1980s involved building an underwater house. This Fricke lived in for eighteen days non-stop at a depth of eleven metres. It was a twenty-six-tonne structure with a wet and a dry room, into which air was pumped from shore. As well as observing new fish behaviour, and taking some wonderful photographs for *Geo* magazine, he experimented with putting free-floating objects into the water, and watched as fishes began to congregate around them. 'This was the first FAD – Fish Attracting Device – which other people later commercialised, and which were used to great effect in the Comoros in the late eighties,' he explained.

Fricke had built the house with a friend, Gerd Helmers, who did most of the steel work. Fricke told him of his desire to take his submersible *Geo* to the Comoros to look for the coelacanth, and of the problems he had encountered in finding a suitable mother ship. Helmers, clearly a good friend, responded with the offer to build Fricke a yacht. The two-masted *Metoka* took longer to construct than *Geo* had, but was finally finished in 1985. With Jürgen Schauer, a colleague of Fricke's at the Max Planck Institute, on board, it sailed from its ship yard in London to Eilat in the Red Sea, where it met up with Fricke and *Geo* for trials. Jürgen then sailed on the three-month journey to the Comoros, where he joined Fricke and Raphael Plante on Christmas Eve 1986, to look for coelacanths for the first time with a submersible.

'It was nerve-wracking,' Fricke conceded. 'I had persuaded the German Research Council and *Geo* magazine to give us large grants for the expedition, but it was by no means a sure thing that we would succeed. It was an expedition into the blue. Of course, I had investigated the possibilities pretty carefully, and I was relatively certain that I would make it, but it was still a risk: if you fail with a research grant on such a scale, then you are lost for the rest of your life.'

The signs were not good. The day after *Metoka* arrived in Grande Comore, the weather turned, and for the next weeks, the expedition had to battle with tropical storms; torrential rain and bucking seas. They were forced to spend days on 'dry' land, sitting in a small café in Moroni, the crumbling capital of the Comoros, gazing at the angry, khaki ocean crashing against the harbour walls. Around them, life continued as normal: unfazed by the seasonal monsoon, Comoran women swathed in brightly patterned cloth sat in the old market in front of small piles of fruit and vegetables, with plastic bags providing an ineffectual shield from the deluge. Children skipped along the narrow alley-ways of the medina, and old men in long white robes played fierce games of dominoes under a broad-leafed mango tree in the *place publique*. Eventually, the rain lightened enough for Fricke to dive for the first time in Moroni harbour, in front of the graceful twelfth-century Friday mosque. Practically the whole city turned out to watch. 'We went down to 200 metres – *Geo*'s maximum depth – and we experienced what became a familiar story over the next three weeks: nothing – a sandy bottom, caves etched from volcanic rock, and very little marine life. It was not dark at that depth, just a little gloomy when the sun was not out.'

They dived at numerous sites along the perimeter of the fifty-seven-kilometre long island, spending many hours beetling around underwater: still nothing. Occasionally, they saw coral heads, with delicate, lattice-work fans fluttering with brightly coloured fish and crustaceans. But most of the time they wended their way along dim and silent canyons, where few creatures chose to live. 'The whole time we were expecting to see Old Fourlegs sitting on the seabed,' Fricke recalls. 'At that stage, everyone thought they rested on their fins, like they were legs. I had a hunch that they didn't walk underwater, but I didn't think that all those famous professors could be wrong. It was just that their fins were not like those

Hans Fricke (right) and Jürgen Schauer inside *Geo*.

of a bottom-dwelling fish. I am not an expert in fish loco-
motion, but I have seen so many fish in my life, and it just
struck me that it was rather fishy!'

At one point, they were diving off the ancient village of
Iconi, when Jürgen Schauer, *Geo*'s pilot, suggested that they
should look into a cave. Fricke, however, demurred: 'A fish
that big would not live in a cave,' he said, so the caves
remained unexplored. At each coastal village, they interviewed fishermen, and were impressed to discover that the
fishermen's estimations of the ocean depth were extremely
accurate. They also learnt that almost all the known coelacanths had been fished up on the west coast of Grande
Comore, and all had been caught at night. After examining
the topography of the ocean bed on both sides of the island,

Fricke came up with the reasonable hypothesis that the coelacanth favoured the west coast because it was structurally
more complex, composed of lava from relatively recent volcanic eruptions. He decided to concentrate their efforts on
the western side.

'As the weeks passed, we became more desperate to see the
fish. We were diving really intensively, even through a cyclone,
which threatened to destroy everything. It was as if we were
involved in a constant fight against the sea.' Fricke's time ran
out, and he was forced to return to Europe having made forty
dives without seeing a coelacanth. Jürgen Schauer was due to
stay behind to continue diving for another five days. They had
decided to dive only at night, since that was when the fish were
caught, and on the night that Fricke left – as he was sitting at
the airport waiting for his delayed flight – Jürgen and a young
student called Olaf, dived off the coast near the village of
Singani, site of Karthala's last eruption in 1977.

'I had a strong feeling about the area,' he recalls. 'It looked
very mysterious – like the fish. We descended at 8.30 p.m. and
at 9 p.m., we saw the first coelacanth, *der Quastenflosser* at the
edge of the spotlight. I held my breath. It was a very, very
exciting moment, one of those special moments of a lifetime.'
He was captivated by the fish's large, shining eyes, its wide
mouth, and its delicate dorsal fin, splayed like an antique
paper fan. 'It was now crucial to see how it would behave
when faced by a large yellow submersible,' he continued. 'We
approached it very slowly, then all of a sudden, right outside
the window, it went into a head stand – just stayed suspended,
nose down, waving its fins slowly and gracefully in the water,
almost as if in slow motion. It was so beautiful, like a ballerina
dancing.'

Unfortunately, the video camera exploded in a puff of smoke when he tried to switch it on, so he took still photographs. After about twenty minutes, the current carried *Geo* too close to the dancing fish, cornering it against the rocks. With a quick movement, it squeezed past the submersible and vanished into blackness. As it knocked the metal, one of its scales came off, which Jürgen managed to retrieve with the manipulator. The submersible surfaced and he jumped into the water and slipped the scale into his pocket before hauling himself on to *Metoka*. 'The mood was very negative on the ship. The captain and crew had almost given up hope and were just going through the motions. Olaf and I decided to keep quiet about it for a bit, so we just got on with things and pretended nothing special had happened. After about half an hour I grabbed the skipper and said, "We got it!" He didn't believe me, and of course, we had no film footage to prove it, so I took the scale out and showed it to him. I still have that scale in my treasure box.'

The next day, Jürgen managed to contact Hans Fricke in Germany who, not surprisingly, was delighted to hear the news. 'Actually, I was more relieved than anything. The operation was saved and now we would get the money for a second expedition. It didn't matter that I hadn't seen it myself: I was just really happy that we had found one.' Over the succeeding nights, Jürgen saw two more coelacanths, which he managed to film with a borrowed camera. 'When I saw Jürgen's footage for the first time,' Fricke relates, 'a big stone fell out of my heart. It was great, really fantastic.

'I suppose, if I was to be honest, I would have to say that the search for the coelacanth was foremost an adventure and a challenge. I knew they must be down there somewhere, and it was fun and exciting to look for them. To find them and succeed in something scientists have been trying to do for half a century was a thrill, a real thrill.'

The team returned in April 1987, and again encountered foul weather. They dived only at night, when the conditions were at their worst: huge swells and high waves crashed against the side of the ship, making the manoeuvring of *Geo* into and out of the water difficult and dangerous. They spent weeks moving slowly through the black water, between steep and barren lava cliffs. There Fricke saw his first coelacanth, a small, near-perfect fish that he christened Nico, and calls his 'son'. Over the next weeks, they watched and followed three more fish – shadowing one for six hours. Fricke was fascinated to study the coelacanth's movement: at first glimpse it looks as if it is waving its many fins in an uncoordinated fashion. But on closer scrutiny, he realised that the fish moves its fins diagonally – left front fin with back right – a gait very similar to a horse trotting, or a lizard walking. He also exploded one of the most enduring coelacanth myths: that it uses its lobed fins as feet, to walk on the seabed. 'Though we observed several individuals resting with their fins braced against the sea bottom,' he wrote in the *National Geographic* magazine, after returning from his 1987 trip, 'we never saw any of them walk, and it appears the fish is unable to do so.' Old Fourlegs, it seems, was not so aptly named.

They had gathered enough footage to make a magical film, which showed the coelacanth, at last, in its natural habitat, swimming with a grace unexpected in a fish so large and – outwardly at least – so cumbersome. It looks and moves unlike any other fish: sculling its splayed fins like a Japanese fan dancer; standing on its head like a gymnast. 'I always say it

'Nico', the first photograph of a live coelacanth.

is a creature that doesn't belong in our marine world,' said Fricke. 'It is a very special fish.'

His film and observations, however, only increased the coelacanth's mystique, raising as many questions as they answered. Where did it spend the day? Why did it stand on its head? How did it feed and reproduce? How many coelacanths were there in the Comoros, and could they live anywhere else? Fricke knew he would have to return if he was going to try to answer these questions. He suspected that the fish spent their days in deeper water, and if this was the case, he would not be able to see them in *Geo*. He determined to build another submersible, capable of descending to greater depths. As soon as he got back to Germany, he started to draw up designs, but it would be two long years before he could return to the Comoros to resume his quest.

In 1988, Fricke took up an invitation from Margaret Smith to deliver a lecture on the coelacanth in Grahamstown. Since conceiving his expedition to dive for coelacanths, he had been in close contact with Margaret Smith and her team at the Institute of Ichthyology in Grahamstown, still a scientific hub of the coelacanth world. Having read *Old Fourlegs* as a boy, he was keen to meet JLB's widow in person for the first time. In the two decades since her husband's death, Margaret Smith had become famous in her own right.

'I have always been one to count my blessings,' she once wrote. 'I realised what a tremendous privilege it was to share the life and work of one of South Africa's great men – realised it every day of the twenty-nine and three-quarter years of our marriage without any regrets. I knew that with the great difference in our ages (nineteen years) I would probably be left a widow; when I married him he was given five years to live – we have had nearly thirty, and few couples can have had a more interesting and productive life than ours.'

She seamlessly assumed her late husband's mantle and responsibilities. People who knew her both before and after his death, talk of a visible 'blossoming'. She cut off her severe bun, and moved from their small, rather awkward house, into a lovely old colonial manor, which she shared with her sister, Flora Sholto-Douglas (known as Aunt Flo, and heartily disapproved of by JLB). Her first goal was to consolidate the J. L. B. Smith Institute of Ichthyology. As she wrote to JLB's estranged sister, Gladys, soon after his death, 'Len refused to have a new building put up for the institute while he was still alive. He preferred to spend the time on his research work . . . And now he has not left me – I am continuing with his work as we had always planned I should. For twenty-two years I worked day and night at his side, and I hope to finish what we started before I die.'

She visited similar institutes around the world, and returned to Grahamstown to work closely with the architect to ensure the establishment which bore her husband's name would be perfect. She became its first director, and her energy and perseverance, together with the Smith name and the coelacanth's fame, ensured she was able to raise the requisite funding. In later years, when the institute was thriving, she hustled for it to be credited as a national museum, with all the financial benefits that entailed. At its head, she quickly became a figure of great stature in her own right: a world-class ichthyologist, an extremely talented illustrator of fishes – over 2,000 accurate and intricate drawings and paintings – and an international ambassador of science.

Margaret Smith was a larger-than-life figure – in both her professional and private life. She took up singing and music again, and she and Aunt Flo held hysterical joke parties, where the humour at times became embarrassingly ribald, reported Mike Bruton – who succeeded Margaret Smith as Director of the J. L. B. Smith Institute: 'It was almost as though she needed to chip away at the academic façade.'

At work, she was adored by her 'children', and her office was perennially crowded with people, as well as donkey harnesses and boxes of stamps for the Red Cross, something JLB, with his exaggerated work ethic, would never have tolerated. She took to committee life like one of her fish to water, and was often to be seen scurrying between meetings, clutching her knitting. 'She would readily end an important meeting to cuddle the baby of a visitor,' Bruton recalls, 'but she was a formidable foe to any government official or academic who criticised her management style. To her, happiness and harmony were of paramount importance. Tea was a loud and joyous occasion as she recounted the hilarious happenings on her early expeditions.'

'Life was always exciting when one was around Margaret Smith,' her secretary, Jean Pote, remembers. 'She was always fired up and enthusiastic about everything, and it seemed to rub off on all those around her. The small things in life were very important to her: if, for instance, she heard of a sheep-dog show in the countryside somewhere, she would sweep off, taking along a handful of her staff for the outing. She gave a huge amount to charity, mostly on the quiet. Beggars would come off the street and she would interrupt a meeting to lend them money, which she would write down in a little black book – always failing, however, to chase it up if they didn't repay her. She made opportunities for people. Especially for underdogs.'

Her generosity was repaid in kind. When her well-known grey Ford was stolen and taken to the nearby township, the thief immediately returned it when he learnt it was hers. And when she got a speeding ticket in Somerset Street, right outside the institute, the cop tore it up when he saw who the driver was. According to her son, William, 'Mom was a saint, the only person I know who could get into a lift full of strangers on the ground floor, and by the third floor, she would know their names and family history and they would all be friends. It was like she developed to be a foil for Dad – and she paid a price for it. When she was growing up and at university, she was tremendously career-minded, but when Dad was alive she had to play second fiddle to his greatness. You can't change nappies and be a great scientist: she was very good at nappies. But she was also great, and she got her greatness through people.'

Margaret Smith embraced life with the same boundless energy as JLB, but without the accompanying grumbles. She continued to go on fish-collecting expeditions for as long as her health permitted. According to Robin Stobbs, who accompanied her on her last trip to Mozambique, she never flagged: 'The endless long days of collecting under the tropical sun, sorting, labelling, pinning out fishes, photographing some and drawing others, preparing for the next day, often only crawling to bed well after midnight, was a routine that Margaret followed with the unflagging enthusiasm which might have daunted someone half her age.' At the age of sixty, she learnt to scuba dive in Hawaii.

She became a fully fledged professor in 1980, and two years later, retired as director of the Smith Institute to concentrate on her last great project: *Smith's Sea Fishes*. This was to be a monumental and definitive tome, successor to the earlier *Sea Fishes* books, which she co-edited with Phil Heemstra, incorporating the work of seventy-two scientists from fifteen countries.

By the time the project began, her health, once so reliable, had already started to fail. A bad attack of arthritis resulted in an operation to have one of her knee joints replaced. In 1985 she contracted pneumonia, septicaemia, and bacterial meningitis. She was in a coma for several hours and not expected to live. However, as Phil Heemstra recalls, 'her indomitable will and determination to finish the book pulled her through. She just refused to die. And even after she was confined to a wheelchair, she continued to be an active presence, driving herself around in an automatic car, lifting herself painfully out of the chair.'

Margaret Smith painting fish for *Sea Fishes of Southern Africa*.

Shortly after the book was finished, Margaret Smith contracted leukaemia. She was in hospital, and in extreme pain,
when Hans Fricke arrived in Grahamstown to deliver his
coelacanth lecture. It was a great success: he showed his new
footage, and was amazed and gratified that some of the audience cried when they saw the film. The next day, Fricke went
to see Margaret Smith in the Port Elizabeth Provincial
Hospital in Port Elizabeth, and to introduce her to the living,

swimming coelacanth. 'It was the only time I met her,' Fricke recalls, 'but she made a very strong impression on me. I thought she was a remarkable person. With her personality and her knowledge, to have lived with such an egocentric husband must have been terrible.'

He projected the first graceful images of the swimming coelacanth on to the bare white wall facing her bed. She was enthralled, and by the time it was finished, she was in tears. 'She said that seeing the live fish had completed the circle of her life, and she was now ready to die,' Fricke remembers. 'She had been very depressed about her illness, but seeing the coelacanth made her incredibly animated and revitalised, emotional and fired up about it. She said she would take the memory to JLB.'

Hans Fricke returned to his submersible blueprints, and six weeks later, Margaret Smith died.

X

JAGO

The new submersible was built in a small wooden shed at the edge of the Max Planck Institute car park. Hans Fricke had decided that, with all they had learnt from *Geo*, he and Jürgen could build it on their own. In order to be able to descend to greater depths, the new machine would have to be more sophisticated: stronger and safer and, inevitably, more expensive. By doing it themselves they would save a fortune.

As with *Geo*, a dummy was built, and gradually the body of the machine was created around it. 'If you do everything on your own, with a small group of people, you rapidly become an expert,' Fricke explained. 'This is what happened to us.' Quite by chance, he met a submersible builder on the road outside his house one day, who agreed to help with the technical specifications. Soon the new improved yellow submarine was ready.

Jago was named after a deep-sea shark with the same colour eyes as the coelacanth, which lives at 400 metres – the maximum depth *Jago* was designed to withstand. It is slightly larger than *Geo* – a metre and a half wide, and two and a half metres long – with bigger windows. Like *Geo*, it is free swimming, fitted with powerful spotlights, an efficient manipulator and a radio communication system with the surface. Fricke took his new submersible for a test dive in Lake Geneva with Jacques Piccard and his submersible, *Forel*, on standby *Jago* passed its tests with flying colours. Fricke was delighted, and took the team to the Red Sea, and from there, back to the Comoros. They arrived towards the end of 1989, with the express intention of finding – and filming – a coelacanth by day.

It did not happen immediately. 'In our imagination, we were sure we would find coelacanths straight away below *Geo*'s 200-metre limit,' said Karen Hissmann, who joined Fricke's team in 1988. From scientific analysis of the coelacanth's blood, it appeared that its best living conditions would be at temperatures of fifteen to eighteen degrees centigrade, occurring below 200 metres in the Comoros, and catch records seemed to tally. 'Our first dives down to 400 metres opened up a new world to us: giant white, funnel-like or branched glass sponges, bizarre, beautiful, fascinating formations – but no coelacanths.'

They noticed how the underwater landscape changed as they descended. Below 200 metres, the dramatic canyons gave way to shallower slopes of ash-coloured sand. They encountered new and interesting fish, and at 400 metres, they found themselves in a dusky, barren landscape. They calculated that at that depth, they were resisting pressure of 3,600 tonnes: 'This incredible thought reminded us of the extreme living conditions surrounding us,' Karen Hissmann continued. 'After a long search we were forced to admit that this wasn't coelacanth country.'

They decided to go back to the area where they had

seen coelacanths two years earlier, and on 5 November at
9.45 a.m., their efforts were rewarded when they caught sight
of a coelacanth standing on its head, silhouetted against the
entrance of a cave. For the first time, Fricke was able to see
the graceful fish in daylight. 'As we approached, it retreated
into the cave,' he recounted. 'We peered inside, and to our
great excitement, there were three more pairs of large eyes,
glowing in the half-light. We saw that the coelacanths were
resting quietly together, close but not touching. At that
moment, we realised where they spent their days, and why we
hadn't seen them before. And the irony was that the cave was
at a depth of 196 metres – within the diving limits of *Geo*.'

Now they knew what to look for, they quickly found many
more coelacanths, sheltering in caves, protected from predators and strong daytime currents. There were never the same
number in each cave: sometimes the fish were in groups of up
to ten; at other times there would be a solitary individual.
They saw three old friends from the 1987 expedition, which
they recognised from their distinct white markings, and noted
one in six different locations. They spent hours upon end in
their yellow steel capsule, just watching. They soon began to
get a picture of the coelacanths' daily routine: resting hidden
by day, then, shortly before dusk, leaving the safety of their
communal caves to forage singly for prey. The coelacanth, it
appeared, hunts alone.

In order to get a better idea of the fishes' movements, a
small 'pinger' – a radio transmitter – was shot, one each, into
the sides of eleven coelacanths, which could then be tracked
from the submersible or the surface. Fricke and his team took
turns to shadow the large creatures as they drifted along, riding the currents like albatrosses in the air. The coelacanths

appeared to be equally happy swimming forwards or backwards, standing on their heads, or lying on their backs – their fat-filled swim bladders enabling them to maintain neutral buoyancy in any position, hovering easily, weightlessly, in the water with only the slightest gentle movement of their fins.

The head stands puzzled Fricke – few fish are known to perform them, certainly not to maintain the position for several minutes without apparent effort, 'as if they were all auditioning for a job in the circus!' Each time the submersible approached, the coelacanth would slowly lift itself into the vertical position, gently waving the tiny central lobe of its tail from side to side. Fricke became convinced that this behaviour was somehow connected with the coelacanth's ability to locate prey by detecting changes in the surrounding electric field – using the rostral organ (the jelly-filled cavities in the snout believed to act as an electro-detection system). He decided to experiment by emitting weak electric fields – mimicking an approaching fish – in the water. Sure enough, as soon as the coelacanth picked up the currents, it swivelled on to its head. While the results were not conclusive, they were plausible. Recently, Fricke identified electricity-generating organs in the tail fin which appear to add weight to his theory: the head stand position, perhaps, enables the coelacanth to detect prey with greater ease.

They photographed every coelacanth they saw from both sides, christened them (Nico was the first to be named), and produced an Identikit for each individual, clearly showing its unique pattern of white splodges on the dark scales. At last count, they had records of 108 different coelacanths. 'Each time we go back, we dread not finding one of our particular

A coelacanth stands on its head.

friends,' said Karen. 'When we see a dead fish we hope we will not recognise it.'

They conducted a thorough survey of the coastal region, in an attempt to estimate the coelacanth population size, and to map their caves. It was not a quick nor an easy task. Coelacanth caves tend to have relatively small entrance passages, which arch into larger internal wombs. Only by lying on the floor of the submersible, and looking through the upper edge of the curved front window, could they watch and film the beasts at home. Even if the entrances were larger, the coelacanths would be hard to see: the spots of white on steel-grey scales melt into the mollusc-studded lava caves, providing near-perfect camouflage. 'It is very, very difficult to see a coelacanth underwater,' Fricke explained. 'If you are not acquainted with coelacanth caves you will never see one. And you need an extraordinarily small submersible.'

On an early visit, in 1987, Fricke and Raphael Plante encountered an expedition team from the Smith Institute, led by Mike Bruton. They met up one evening – along with Guelph University's Eugene Balon – in a grimy, flea-bitten restaurant in Moroni. By that time, over 140 coelacanths were known to have been caught since 1952, and by their estimates, there could be as few as several hundred remaining. As they discussed their separate findings, they realised that there was going to have to be an international effort if the coelacanth was to be 'saved'. They set up the Coelacanth Conservation Council (CCC), with the aim of gathering as much information as they could, and making it available to as wide an audience as possible. This, they hoped, would stimulate interest in the coelacanth, leading to a release of funds.

The principals of the CCC embarked upon a worldwide

lecture tour to raise awareness for the plight of the coelacanth. 'All the time, we bashed heads against the furry-
mammal brigade,' Mike Bruton complains. 'It is much easier
to raise money for the soft and fluffies than for the wet and
slimies, especially if, like the coelacanth, they live so deep and
can't be seen in their natural environment, or even in an
aquarium.'

They didn't have much success in raising money, but in
1989, after strong lobbying by Fricke, the Convention on
International Trade in Endangered Species (CITES) slapped
its strongest protection order on the coelacanth. It was listed
under Appendix One (along with blue whales, snow leopards
and Sumatran tigers), which declares it to be a severely endangered species and strictly forbids any trade in coelacanths.
This was considered to be a tremendous achievement – if it
could be enforced.

President Abdallah, however, for reasons of his own,
refused to sign the CITES accord, leaving the door open for
the Comoros to export coelacanths, dead and alive. Only a
few months after the coelacanth was elevated to Appendix
One, the *Jago* team ran into a head-to-head confrontation with
the multi-million-dollar expedition from the Japanese Toba
Aquarium, intent on trapping a live coelacanth. Fricke has
always been vehemently opposed to the concept of having
a live coelacanth on public display – even if it were proven
to be easy and safe to catch them. Before embarking on the
1989 expedition, he had protested against Toba's project, and
succeeded in provoking President Abdallah to issue a decree
prohibiting the export of live coelacanths. But without his
formal signature on the CITES accord, the Japanese paid little
heed. They continued to lower their steel-mesh catch cages.

Eventually, Fricke snapped. After a series of underwater encounters with the vicious-looking cages, he decided to take action. He prepared two laminated cards: one with a picture of the coelacanth above the legend, 'Coelacanths – Let them where they are!' The second identified *Jago* as the sender. One night, they took *Jago* to a cage, and with the manipulator, carefully hooked the cards into the mesh. Shortly afterwards, the unsuccessful Toba team was ordered home – apparently at the express wish of the Japanese Emperor.

'We need a live coelacanth in captivity,' said Mike Bruton, who left the Smith Institute to work at the Two Oceans Aquarium in Cape Town. 'There are some vital gaps in our knowledge of population dynamics; their age and size at maturity, number of young produced, gestation period, growth rate, final age – some of which could only be obtained from observation of captive specimens.' He believes a captive coelacanth would raise public awareness of a rare and beautiful beast, which would lead to an injection of funding for research, and conservation activities in the Comoros. 'Once we know how to catch them, then we could increase the number in captivity, and perhaps breed from them. I am sure that we could easily get one, with the help of Fricke and his submersible. But he refuses to co-operate.'

Hans Fricke has steadfastly maintained his opposition: 'It is a mad idea!' he exclaimed. 'Nobody has seen a juvenile. Nobody knows about the sexes. Nobody knows how they do love, and under what circumstances. And they think about putting one in an aquarium to breed? It is ridiculous! A South African aquarium already claims to have the historical right to be the first to put a coelacanth on display – but this is nonsense: nobody has the historical right to a fish! And if one aquarium gets its coelacanth – then what? Everybody else wants one, creating a run on the small coelacanth population. No, I will not help!'

It is the small – and, he believes, diminishing – size of the coelacanth population that worries Fricke most. The more he finds out about the coelacanth and its habitat, the more he despairs of its continued survival. Almost constantly, the fish is being subjected not only to harassment by scientists from around the world, keen for their own specimen, but also to increased fishing activity by the growing Comoran population. Each time they have been to the islands, Fricke and his team have conducted a rough census. Between 1989 and 1991, the number of coelacanths they counted in the census area appeared to remain fairly stable, indicating a total population around Grande Comore of somewhere below 650. By 1994, however, the number appeared to have declined drastically, a result, Fricke believes, of increased fishing pressure by the economically stressed fishermen.

There are some coelacanth scientists, however, who regard Fricke's population numbers as unduly pessimistic. Robin Stobbs believes the population could be many times greater than Fricke's estimates. 'He has concentrated his calculations on the areas where the fish are caught, without really looking elsewhere,' Stobbs maintains. 'But if you examine it in terms of fishing methods, you see that, for instance on the east coast of Grande Comore and on Moheli, Mayotte and Madagascar, they don't fish in the same way, so they are never going to catch a coelacanth. And, although there has been an increase in the total number of fishing *galawa* in recent years, there has actually been a reduction in night-fishing activities, which is good news for *Latimeria*. I believe there could be many more fish out there, in no danger of being caught at present. After all, we don't yet know the ancestral home of the coelacanth: where was it from the time the fossil records disappeared seventy million years ago, until the Comoros were created sixty million years later? Surely not paddling about over the deep ocean waiting for a volcanic island to pop up from below!'

'If you examine the socio-economic situation in the Comoros,' explains Fricke, 'you see why the Comorans have to eat every little creature from the sea. The reef is virtually fished out, and they are having to go after deeper water resources. While they do that they catch the coelacanth by chance. Who can blame the poor devils? Their survival is more important than the survival of a fish. People have to eat. It comes from our aggressive way of handling our planet and really it is very sad, but this is the way evolution works, and unfortunately at this time we are a constraint which is terrible.'

In 1994 – five years after it was promulgated – the Comoros signed the CITES accord, whereupon the official market in coelacanths collapsed. It is suspected, however, that there is still a black market, at least when the right buyers are in town. Some reports estimate that coelacanths can still fetch up to $2,000 – five times the annual average salary of a fisherman, and very hard to resist. But, as many Comoran fishermen testify, catching a coelacanth these days is no longer an automatic lottery win.

Ahmed Bourhane, a fine-featured, shaven-headed fisherman from the village of Mindradou, remembers only too well the time he caught his *gombessa*. 'It was 1995,' he recounts. 'At about 11 p.m., I felt something take the bait, but at first I didn't realise what it was. When it came up, it was moving like a woman in bed. It was very heavy, and I only managed to put a big hook through its mouth on my third attempt. I touched its skin and realised it was a strange fish, but it was dark and I didn't know it was *gombessa*. I called to the other fishermen who had lights, and when they came over, I realised what it was.

'Oh, I was very happy because I believed I was going to be able to do my *grand mariage* and go to Mecca: before, when fishermen from my village caught a *gombessa*, they sold it for a lot of money. I put it in my boat, and did everything I could to keep it alive. It didn't struggle.' It was Sunday and there were no taxis, so he hired a car and took his fish to the museum in the capital, Moroni. They said they didn't want it. He took it up to the Galawa Beach Hotel on the northern tip of the island, but they didn't want it either. Nor did the Chinese Embassy. 'I had the fish in the car the whole day and spent 40,000 CFA (about £57), but no one would buy it. So I drained the oil, which I give to people when they are sick, and grilled the fish. It didn't taste too bad.'

The further that word of experiences like that spreads – and news travels fast in the Comoros – the better the coelacanth's chances of survival. Even Hans Fricke is encouraged by an increasingly organised grass-roots conservation movement in the Comoros, which employs somewhat unorthodox methods to get its message across. The *Association pour la Protection du Gombessa* was formed in 1994, and formally inaugurated four years later, at a colourful ceremony in a whitewashed class-room, seemingly stranded in the middle of a lava-rock plain. Fishermen and elders from twelve villages, dressed in their best clothes, sat straight backed behind old wooden desks;

outside, women draped in brightly patterned cloth danced in circles, ululated and sang about the coelacanth. '*Gombessa* is a wonderful fish,' they sang. 'The world likes *gombessa* – it is better to protect it and not to fish it, in order for our children to know it.'

The proud President of the Association, a veteran fisherman called Hassane Djambaé opened the proceedings. Beneath his crimson fez, his face was a study of commitment. 'We are here today to talk about how to protect *gombessa*,' he proclaimed. 'It will not be easy. Fishing is our culture, and this is one time when we have to teach fishermen not to fish.' He was followed by a young boy who sang verses from the Koran.

'The aim of the association is to protect the coelacanth and also the bottom of the ocean,' explained Said Ahamada, a young Comoran ecologist and member of Ulanga, the national environmental conservation group which has been a strong impetus behind the formation of the Coelacanth Protection Association. 'Comoran fishermen do not make much money – on average about 2,000 CFA (£2.80) a day. They need help. It is hard for them to understand the global significance of the coelacanth, and so we stress the socio-economic benefits they might accrue from its conservation.'

The ultimate goal is to get the waters around Itsoundzou – where the highest concentration of coelacanths are thought to live – declared a National Marine Park, with enforced prohibition of fishing. Hans Fricke wants to plant a permanent camera at the mouth of a coelacanth cave, to transmit live footage to a coelacanth information centre on the shore. 'It must be something exciting, something extraordinary,' he stresses. 'You can't touch the animal; you can't dive to it; you can't have a tourist submarine because it would be too expensive, and wouldn't be able to get close enough. So the only way to see a live coelacanth is through real-time video images.' He is convinced that the Comoran tourist trade would make the centre economically viable. All that is needed is the money to fund it.

The Coelacanth Protection Association is pushing this plan, telling the fishermen that if they leave the coelacanth alone now, they will benefit in the future. Some enthusiasts have promised dire consequences to those who bring back a coelacanth; the news is spreading that far from receiving fame and riches from the *gombessa*, the fisherman who catches one will instead be cursed by a witch-doctor. Already, several fishermen have reportedly cut loose a coelacanth, rather than bring it to shore. Fricke thinks this the best news he has heard: 'But, but, but ... Can the coelacanth survive being caught and released? They are pulled up through the warm water, totally distressed and exhausted, there is a build-up of lactic acid, damage to the inner ear, and the respiratory system is severely stressed. I fear that the fish will have no chance. But maybe they are tougher than we think? Maybe if they are brought to the surface and immediately released, they will survive?'

Jerry Hamlin of the Explorer's Club is convinced of it, and has staked his own money and energy as proof. After the infamous Mombassa was killed, and Jerry discovered that his resuscitation tank had been dismantled, he decided to try a different tack in his ongoing battle to save the coelacanth. From the headquarters of the Coelacanth Rescue Mission, a picturesque former inn in wooded Greenwich, Connecticut, which he shares with an albino Burmese python, a Colombian

red-tailed boa, an Indian Star tortoise, a Parson's chameleon
and a bald cockatoo, he set up a prize-winning coelacanth
website. 'I ran a "Save the Coelacanth" contest on the
dinofish.com website, offering a $500 prize,' he recounted.
'For months I got the expected suggestions of cloning, fish
farms, artificial reefs, and so on – all hopelessly expensive
and impractical. Then one day, I got an e-mail from a Dr
Raymond Waldner, a biology professor in Florida. He mentioned a technique that was used to release deep-reef fish
without causing them the fatal stress of having to swim back
down: a barbless hook with a weight attached to the shank is
inserted upside down in the fish's lip. A line is attached to the
curve of the hook and the fish lowered by that line to the bottom. When the line is jerked up, the barbless hook pulls out
of the fish's mouth. The fish is free on the bottom and the rig
is pulled back up to the surface where it can be used again.'

Hamlin read the e-mail with interest, and tried to work out
how it could be applied in the Comoros. 'Then the coin
dropped. I devised a way to make the rig extremely light and
small, so light and small that it could be placed in a pouch and
sewn to a T-shirt. The T-shirt could then be distributed to
fishermen, and instructions for the system could be printed
iconically on the back of the shirt. What an advantage:
because this way the coelacanth's time in the warm water of
the ocean surface could be reduced to a few minutes. Dr
Waldner got the $500. Then I set about designing the shirt,
manufacturing the rigs and sewing on the pouches. We produced about seventy, and the first batch was airmailed to the
Comoros, at a cost of over $1,000. The T-shirts were distrib-
uted to the fishermen of Itsoundzou in August 1998 by Said
Ahamada, while hundreds of Internet contributors bought

rigless T-shirts over the web, helping to fund the project.' If this method proves to be successful, he plans to send small fish tags to the islands, which could be used to mark released fish, so that they would be noticed on recapture, or observed from a submersible, indicating that the fish did recover. 'I just hope this works and the coelacanth is saved, so that I can get on with my life,' he said.

In 1991, an event had occurred which gave hope to some of those who feared the coelacanth population was in danger of extinction. Some time in August, a huge female coelacanth was caught in the trawl net of a Japanese ship off the coast of Mozambique. The 1.79 metre-long, ninety-eight-kilogram fish was frozen on board, and in December 1991, handed over to the Natural History Museum in the capital, Maputo.

True to form, it was on Christmas Eve that the J. L. B. Smith Institute in Grahamstown received a fax informing them of the coelacanth's capture, and a few weeks later, Mike Bruton (at that time still director) and Hans Fricke went to Mozambique to investigate. Bruton was met at the airport by the museum's director, Dr Augusto Cabral, an extraordinary character who had single-handedly kept the museum going throughout his country's long civil war. Cabral reassured them that it was definitely a coelacanth – only the second to have been trawled, and indeed to have been found anywhere but the Comoros. The bad news was that – in another uncanny mirror to the events of 1938 – lacking the facilities to keep it frozen, Dr Cabral had been forced to dissect the fish and discard its internal organs. The good news, however, was that he had found twenty-six perfectly formed coelacanth pups inside, which he had managed to preserve.

Islands of the south-western Indian Ocean

The Mozambique fish threw a slew of theories and assumptions into the air. Not only did the place and method of its capture revive former conjectures that the coelacanth inhabited a much wider geographical area – and that the East London specimen was not necessarily a stray – but it also threw oil into the simmering debate about the coelacanth's reproductive processes.

Until that point, all estimates of coelacanth population dynamics had been based on the American Museum's specimen, with its five embryos. Then along came the Mozambique fish, quintupling the potential birth rate; if coelacanths could produce up to twenty-six young, then perhaps they weren't as endangered as we thought? It also destroyed, once and for all, Eugene Balon's theory of inter-uterine cannibalism.

To the sceptics and pessimists of the coelacanth world, the Mozambique specimen wasn't such a big deal. Any live bearer – even one that produces twenty-six offspring – is a slow reproducer, and it was entirely possible, they argued, that, like the East London fish, this coelacanth had also got caught in the southerly current and drifted from the Comoros to the waters off Pebane where it was caught.

There had long been rumours that coelacanths were to be found off Madagascar. As far back as 1982, the government issued a miniature sheet of stamps depicting the coelacanth in its natural habitat. Four years later, a specimen purported to have been caught locally was displayed in Tamatave (now Toamasina), although it was later believed to have been bought in the Comoros. Jerry Hamlin spent several weeks travelling the length of the north-eastern coast, interviewing

The Mozambique coelacanth pups, from a litter of twenty-six.

fishermen, but was unable to come up with concrete proof of a Madagascan coelacanth. The government, however, seemed fairly sure, and in 1993, issued a new coelacanth stamp – this time with a background of rocky overhangs and caves.

Two years later, their faith was justified. On 5 August, a thirty-two kilogram coelacanth was hauled up in a deep-set shark net off the village of Anakaó, thirty-three kilometres south of Tulear on the south-west coast of Madagascar, and

1,300 kilometres south of Grande Comore. It was caught by three young fishermen, none of whom had ever seen a coelacanth before. When they pulled up the net and saw their unusual catch, they were terrified and nearly threw it straight back in the water. The Malagasy follow a complicated system of taboos – or *fady* – which relate to their belief in mysticism and the spirit world, and which are rooted in their Indonesian ancestry. Anything that is strange or unknown is automatically *fady*, and it is prohibited to touch or handle it. To the Malagasy fishermen the coelacanth, with its strange 'limbs', must have seemed like a freak of nature, and as such to be treated with extreme caution. Breaking a *fady* is a potentially fatal business. Nevertheless the owner of the boat, ZeZe, decided to risk it and take the coelacanth back to the village to show to the mayor, Mr Regis Robinson, an experienced fisherman.

Regis Robinson, however, was just as confused, and the fishermen were on the point of cutting up the coelacanth to use as bait, when a passing Frenchman recognised it as *notre coelacanthe*, and bought it for 20,000 Franc Malagasy (Fmg) (£4). He took it in his boat to the Museum of the Institute of Fisheries and Marine Sciences in Tulear, where it was preserved and exhibited.

Perhaps inevitably, this was not enough for everyone. Some sceptics suggested that this too was a stray, carried on the currents from the Comoros. But when, in 1997, another specimen was caught, apparently by the same fishermen, in the same location and under the same circumstances, the coincidence was too great.

'It could very well be that Madagascar is the ancestral home of the coelacanth,' said Robin Stobbs. The huge island – fourth largest in the world – was once part of the supercontinent of Gondwanaland, before splitting from the African mainland around sixty to seventy million years ago – at about the same time as fossil evidence of the coelacanth disappeared, making it far more ancient than the relatively new Comoros. 'It was only very recently that fishing techniques changed in that region of Madagascar, and instead of using deep long lines to catch sharks, the local fishermen started using nets, into which this unfortunate coelacanth blundered,' Stobbs continued. 'Until recently, few fishermen risked dangerous night fishing, far from shore, where they might drift out to sea and die. The bodies of the dead are treated with extreme veneration in Madagascar and the supreme *fady* is for your body to go unburied – this reflects badly on your family and entire community. So perhaps the coelacanth has been there all along, but it has taken a change in fishing techniques to catch one.'

Perhaps the coelacanth has been living quietly even further away, all along? Over the years, a series of clues have emerged which seem to indicate that its home range is wider than believed. Some of these clues are undoubtedly red herrings, and taken alone, none of them has provided conclusive evidence that coelacanths live anywhere other than the western Indian Ocean, but taken together, at the very least, they open a doorway of possibility.

Among the most compelling is the Tampa scale – the unusual fish scale sent in 1949 by a Florida souvenir seller to Dr Isaac Ginsburg at the Smithsonian in Washington. At this point, the only coelacanth known to the world was Marjorie Courtenay-Latimer's, sitting stately and stuffed at the East London Museum, with all its scales accounted for. While the

Tampa scale was not identical, it bore enough similarities to *Latimeria chalumnae* to convince Ginsburg that it was the scale of an ancient fish, probably a crossopterygian, and possibly a species of coelacanth.

Though Ginsburg wrote to the woman, asking for further details, she never replied, and the provenance of the Tampa scale remains a mystery – a tantalising hint at the existence of an American coelacanth.*

The next major clue came to light in 1964, when the Argentinian chemist, Dr Ladislao Reti, bought the first silver *ex voto* from a village church near Bilbao, in Spain. His intricately crafted, four-inch sculpture was identified positively as a species of coelacanth – every feature was identical, down to the engravings on the scales, which clearly represent the coelacanth's identifiable white markings. He took it to an American palaeontologist who, after careful examination, concluded that it was probably of a different genus to *Latimeria*. In appearance and detail – particularly the number of scales – it is more similar to the fossil genus *Macropoma*, the youngest of coelacanth fossils to have been discovered, dating back around seventy million years.

Maurice Steinert's model, which he bought shortly afterwards in Toledo, Spain, is larger than the first *ex voto* – almost exactly the same size as the Mozambique pups – and even more delicately and intricately carved, undeniably a coelacanth. Steinert bought it from a collection of silver fish, probably by the same artist, and all of which depicted unusual fish with exceptional body shapes. In the opinion of an authority on South American silver at the Prado Museum in Madrid, it

*The scale is now lost, presumably somewhere in the vast museum archives.

is probably the work of a seventeenth- or eighteenth-century Meso-American artist. While Spanish silversmiths of that era stamped their work with the date and place of origin, Mayan Indian silversmiths were prohibited from doing so – and neither silver coelacanth had such a mark. It was apparently common in those days for rich Spaniards to bring back works of art from their colonies in Central and South America, and donate them to the Church. Another silver expert who was shown the model by Plante and Fricke confirmed the Prado authority's dating: both the black silver oxides on the surface of the fish, he said, and the minute forged joint beneath the head (later such joints were cast and appeared heavier), were consistent with seventeenth- and eighteenth-century work. It is possible that these beautiful coelacanths are artistic representations of the same, mysterious population of fish which shed the Tampa scale.

Hans Fricke and Raphael Plante were particularly fascinated by the question of how the silver coelacanths could have been crafted several centuries before the 1938 coelacanth was brought to world attention. The representation of the white markings on the scales appeared to preclude a fossil muse, and anyway, the replica is so accurate in every tiny detail, that it would have had to exceed the patient reconstructions of even the most skilful palaeontologist. 'For even the best of artisans to achieve such lifelike perfection from a fossil seems incredible,' wrote Donald de Sylva in *Sea Frontiers*.

Conceivably, the artisan could have been working from a fresh fish found in the Comoros long before it was identified by science, or brought back to Central America from the islands, salted or dried, but that again defies reason.

The easiest explanation is that the artists had live models to work from: fish found nearer to home – perhaps in the waters off Central America. There are many remote ocean areas in the world, with similar environments to the Comoros – rocky, volcanic caves and still, deep waters. Surely it is possible that they might exist elsewhere in the fathomless oceans of the world – but that we don't employ the right methods to ensnare them?

Hans Fricke cannot discount the possibility. 'I don't see why there might not be another population of coelacanths somewhere. I just hope we never find them.'

20°N

10°N

South China Sea

PHILIPPINES

Celebes Sea

Singapore

Manado Tua

Manado

BORNEO

Equator

Jakarta

JAVA   *Bali*

0   100   200   300 miles
0   200   400 km

110°E   120°E

Indonesia

XI

RAJAH LAUT

It was a dream wedding and honeymoon. Twenty-eight friends and relatives flew from around the world to Bali, to witness Mark Erdmann marry Arnaz Mehta in a traditional ceremony. There was a procession of beautiful maidens carrying offerings of fruit and flowers on their heads; a wedding feast eaten from a long table groaning with local delicacies; dancing under a full moon. The next day, the wedding party boarded a tall-masted *phinisi perahu*, for a six-day honeymoon cruise to the Komodo Islands.

At the end of the trip, most of the guests flew home, while Mark and Arnaz and two friends, John and Janel Intihar, continued their holiday in Sulawesi. For the Intihars it was an incredible experience – neither had been to Asia before. Their trip ended in Manado, on the northerly tip of Sulawesi, and on their last day, 18 September 1997, Mark and Arnaz took them on a final, cultural excursion: to see a real, stinking Indonesian fish market.

It was a baking hot morning. As soon as they stepped out of the taxi they were enveloped by the hustle and bustle of the market; people chattering and staring at them, everywhere the stench of fish. Arnaz caught sight of an old, wrinkled man wheeling a wooden cart with a scarlet A painted on the side, across the car park. On it, shimmering in the heat, lay a large and unfamiliar fish. She called to Mark, a marine biologist, and asked if he knew what it was.

'I immediately recognised it as a coelacanth,' he recalls. 'There was no doubt in my mind about it. I had read a book about the coelacanth when I was twelve, and it had really caught my imagination. At that time I knew they had only been found in the western Indian Ocean, but I wasn't up to date: I didn't know whether it had been found in this region since then, and I couldn't really believe that we had just stumbled into a major discovery.'

He explained to the others what the fish was and why it was interesting. His mind was already churning, as he tried to decide what to do; whether to buy the fish then and there. 'I was rather wary of getting too excited about it, as I had been burnt like that before. Apart from the eleven new species of stomatopods (mantis shrimps) that I've described from Indonesia, there have been all sorts of other creatures or types of behaviour that I had thought were possibly new to science. I went to great lengths to examine them, and take photographs – sometimes even to the point of killing and pickling a creature in a jar. But inevitably, when I've taken them to the Smithsonian Institute in Washington or another such institution, I've found out that they were of no particular interest. It has been embarrassing, and I've felt particularly bad about the animals I killed.' The coelacanth, at least, was already dead; but on the other hand, it was far too large to fit in a jar.

Mark remembered that they were only in town for another

few days, staying in a small hotel room, with a huge number of things to do. He and Arnaz were moving to Manado a month later, and had to arrange all the practicalities before they returned to America, so common sense won the day. 'I was really intrigued by it,' Arnaz says. 'I thought we definitely should try to find out more. A large crowd had gathered around us; I don't know what interested them more – us or the fish. They started speculating as to what it was, but they clearly didn't have a clue – most seemed to think it was a deep-water grouper.' Arnaz encouraged Mark to photograph the fish and talk to the old man. He didn't have his own camera so he borrowed Janel's, and quickly took a few pictures of the coelacanth on its wooden cart.

While Mark examined the fish, Arnaz questioned the old man. 'He seemed to be very uncomfortable with the attention,' she recalls. 'He looked as if he would be happier if we left him alone to get on with selling his fish to a market trader. I asked him where he had caught it.

'"*Dasar Laut*," he replied [On the sea floor]. This was not overly informative.

'I asked again, "Where?"

'He turned and pointed out to sea, towards the offshore islands.

'"Far," he said.

'"Do you catch them often?"

'"Rarely."

'His short answers stuck in my mind,' Mark continues. 'As he was so disinclined to elucidate, I broke the first rule of questioning and started to ask leading questions. Anyway, when I pushed him, he said he had caught it in deep water, and seemed to suggest it was while he was hand-lining from his dugout. It was at night, and the fish hadn't wanted to die.'

The old man was becoming increasingly uncomfortable, and while they talked, his fish was baking in the hot morning sun. They decided to let him go about his business, confident that they would be able to recognise him again, by the scarlet A on his cart. 'I remember my hesitation as he was walking away,' says Mark. 'Should I have bought it? Logic kept saying no. Of course, I soon came to regret my decision. I kicked myself about it for nearly a year; it was the biggest mistake I have ever made. I sat up many nights agonising about it: at the very least I could have taken a scale, or blood and tissue samples – I even had my sampling kit with me, but it just didn't occur to me at the time. I consoled myself that we were going to be in Manado for two years, and we were bound to find another one.'

Three days later, they flew to California. During the long flight, the coelacanth kept swimming into Mark's thoughts. The day he got to Berkeley, where he had recently completed his doctoral thesis on stomatopods, he went to see his department head, Dr Roy Caldwell, and asked him if he knew whether coelacanths had ever been found outside the western Indian Ocean.

'Not to my knowledge,' Caldwell replied. They searched the internet and their reference books, but found no mention of coelacanths living within thousands of miles of north Sulawesi. They were both very excited about the apparent significance of the find, but Mark decided to wait until he had Janel's photographs from the market before getting in touch with any coelacanth specialists.

Four days later, he and Arnaz were in Ohio, visiting Mark's mother. They returned to the house after lunch, to find that

Mark Erdmann, in search of the Indonesian coelacanth.

the phone had been ringing off the hook. Roy Caldwell had called several times, as had their friend John Intihar, and David Noakes, an ichthyologist from Guelph University in Canada. All, apparently, were excited about some fish, Mark's mother reported.

Mark contacted Roy, who said he had possibly made a big mistake. He had picked up an e-mail from John Intihar (a systems engineer in Washington DC), which had been sent to everyone who had been at the wedding, urging them to visit 'Mark and Arnaz's Honeymoon Website', which he and Janel had created. Roy, a self-confessed technophile, had immediately hooked up to the site, where he saw – among pages of wedding pictures – a photograph of the coelacanth. There it was, in full colour, definitely and unmistakably Old Fourlegs. Roy's first excited thought was to e-mail some Berkeley colleagues, telling them to sign on and 'see what Erdmann caught in Indonesia'.

Among the people he contacted was George Barlow, a well-known ichthyologist and fish behaviourist, who forwarded the message to a former student of his, David Noakes, a colleague of Eugene Balon's and fellow member of the Coelacanth Conservation Council. Noakes was well aware of the rumours of predatory Japanese aquariums and Chinese life-prolonging elixirs, and as soon as he saw the picture, he thought of the conservation implications. He phoned Roy, to congratulate him on an amazing discovery, but warned him to get the picture off the web straight away. A quick call to John Intihar, and the coelacanth picture disappeared within the hour. But by that time, a number of people had already learnt of the Indonesian fish.

Roy's phone started to ring. There were calls of congratulations but also, inevitably where the coelacanth is concerned, a great degree of scepticism. Among the first to contact him was Eugene Balon, keen to stress that it was without doubt a 'honeymoon hoax'. Vic Springer at the Smithsonian, initially sceptical, later came up with a theory that the coelacanth had most probably been caught by a Japanese trawler in the Comoros, but offloaded in the Manado fish market on the way home out of fear of contravening the CITES regulations.

By the time Mark got back to Berkeley, he had decided not to publish his find straight away. 'I didn't want all sorts of people descending on Manado, and offering the fishermen large sums,' he says. 'I wanted first of all to dampen the scepticism by getting hold of another specimen for myself, then to make sure that conservation measures were in place before announcing it to the world.'

The Erdmanns returned to Manado, a bustling seaside city described by naturalist Alfred Wallace in 1859 as 'one of the prettiest in the East', in the hope of finding another specimen before long. They moved into their house on the island of Bunaken, a fifteen-kilometre boat ride away, and one of the world's best dive sites. It was an idyllic place to live for two years, with magical coral gardens, sandy beaches, no motor cars, and a relaxed, island population. They set about searching for a coelacanth. Their first stop was the fish market in Manado, where they spent days in the hot sun, looking for either the old man, or his scarlet A cart. Of neither was there any sign. The cart, they soon realised to their dismay, could have been repainted – and even if Mark and Arnaz found it, there was no guarantee it would lead them to the old man.

They tried another tack. Mark got reprints of the photographs of the coelacanth made and distributed them around the market, with the offer of 200,000 Rupia (Rp) (£12) compensation to the person who brought him a coelacanth. None of the fish vendors, however, appeared to recognise the fish. Many of them said it was a grouper, or made other implausible suggestions. One man, however, appeared to know it. He said it was called *kabos laut* – roughly translated as mudskipper of the sea – which to Mark's ears sounded a definite possibility.

'By this time the search had really taken hold of me. I read everything I could find about the coelacanth, including *Old Fourlegs*, and the story gripped me from the first page to the last. Since I was a young boy, I have loved stories of marine exploration and discovery, and JLB and the coelacanth adventure really appealed to me. I would have liked to have lived in Darwin's time, when instead of scientific specialists there were true naturalists who travelled the world discovering new and interesting things. So I was determined not to let this opportunity slip by: I was determined to find another coelacanth, even though *Nature* had indicated that they would be willing to publish a report based on the photographs alone. I didn't want simply to publish and kiss the coelacanth goodbye.'

With the aid of a grant from the National Geographic Society, Mark decided to expand the perimeters of his quest. He began to tour the offshore islands, asking fishermen if they knew the coelacanth, and leaving copies of the photograph, with his address and the reward details on the back.

Rp 200,000 – per tail, maximum three fish.
If you catch one, please bring it quickly and directly to Bunaken
and look for Dr Mark Erdmann in Pangilisang Beach.
Please bring it immediately, before it starts to go off.

He started on Bunaken. 'The responses I got fell into three categories,' he explains. 'A blank stare; those who said they knew it as *buku laut* (literally, book of the sea), a big fish which they said shelters under flotsam during the north-west monsoon – I dismissed that as highly dubious; and those who said it was *ikan sede*. The last I found initially promising. They were mainly old hand-liners, who fish from *pirogues* identical to the Comoran *galawa*. They seemed to recognise the important aspects of the fish – the fins, tail and scales – and said they typically found it at a depth of more than 100 metres, near the coral wall. They got very excited about catching it, and said they would have one for me soon. My hopes were raised.'

The next week, he visited another small island, Pulau Nain, where he talked to shark-net fishermen, who said they didn't know the fish. 'You don't know *ikan sede*?' Mark asked.

'Sure, we know *ikan sede*, but this is not it,' they replied.

'We went from high to low – it was an emotion rollercoaster,' Mark recalls. 'Half the time I felt really discouraged, not knowing who to believe and who not to. Anyway, I left everyone with a photograph and my address.'

In the second week of March, Mark and Arnaz visited the next-door island of Manado Tua, a defunct volcano which rises like a huge anthill out of the sea. The islanders live along the narrow fringe of beach; most earn their living from the sea, while a minority harvest coconuts, and farm bananas, mangoes and hot bird's-eye chillies on the densely vegetated steep slopes of the mountain. Mark and Arnaz climbed to the top, leaving their cook, Tante Ita, whose late husband was from Manado Tua, to ask around among the fishermen. By the time they got down again, she was jumping with

Manado Tua, as seen from Manado harbour.

excitement: 'I've found someone!' she cried, and took them to meet Om Lameh Sonathon.

Om Lameh, a slight man with a shy smile, was fifty-six, he told them, and had been fishing for fifteen years. Every evening, he takes his crew of eleven men on his flaking twenty-foot boat, *Trinitas*, to lay deep gill nets near the coral walls. The primary targets are sharks, which have their fins cut off before being sold to China and Taiwan for shark's fin soup, deep-water grouper and snapper. According to Mark, 'He seemed to recognise the picture immediately. Yes, he had caught them – perhaps two or three a year – off the south-eastern coast of Manado Tua. No, it was definitely not *ikan sede* – this fish was thicker and more oily. They called it *rajah laut* – king of the sea.'

It was the first time Mark had felt confident about an

identification. His confidence was further boosted when Om Lameh suggested he talk to another gill-net fisherman, Maxon Haniko, on the west coast of the island. 'Maxon was a younger man, perhaps in his early thirties, with a rather cock-sure manner. He also appeared to know the coelacanth, and was very keen on the prospect of looking for one for me, but unfortunately, he had a problem with his boat engine ... I agreed to lend him one of mine for a while. We went back to our boat to get it, and when I returned to give it to him, there was an old wrinkled man sitting with him, who looked vaguely familiar. Maxon introduced him as his father, and when I sat down and started telling my story again, about the fish in the market, the old man perked up and said excitedly, "I was the man in the market!"'

This was the confirmation Mark was looking for. Om Maxon's father explained that he had been there selling a *rajah laut* caught by Maxon – he himself had never caught one while hand lining from his small *pirogue*. He said he had sold it to a fish trader for Rp 25,000 (at that time, about £6), who had immediately sold it on to a Chinese person. In fact, the old man said, each time he had sold a *rajah laut*, it had been bought by a Chinese person. 'He thought this was really funny,' Mark recalls. 'As a rule the Chinese here are more wealthy and more ostentatious with their wealth, and the fish has a reputation for causing diarrhoea. The old man thought the Chinese bought it believing it was a grouper, and then ate it – to find themselves shitting oil!'

If the coelacanth was *rajah laut*, what then was *ikan sede*? Mark and Arnaz didn't have to wait long to find out. A few days later their boatman, Daeng Said, came to the house early one morning to say that an *ikan sede* had been caught and was being saved for Mark by a fish broker on the beach at Bunaken village. Said didn't think it was the fish they were looking for, but they rushed to the beach nevertheless. There they saw a very large brown fish, with big eyes, a big mouth and short, spiny scales. Mark could see that it was a deep-water fish, but beyond its size, it bore no resemblance to the coelacanth. He thanked the fish broker and paid him a decent sum, took some photographs of the fish, and gave a portion each to Said and Tante Ita to try. Both were a little hesitant at first – they had heard the stories of oily diarrhoea – but Tante Ita ate it, and pronounced it delicious.

Mark went home to try to identify *ikan sede*. He found a picture of the same fish on the US Food and Drug Administration website. It turned out to be the coelacanth's old friend and Comoran neighbour, *nessa* – the oil fish, *Ruvettus pretiosus*. 'We now knew that the Bunaken fishermen clearly didn't know what they were looking for – the tails of the coelacanth and the oil fish couldn't have been more different. We decided to concentrate our efforts on the Manado Tua crews.'

Mark started visiting Om Lameh and Maxon every couple of days. He raised the reward to Rp 600,000, roughly equal to the market price of two large sharks. 'I thought carefully about the amount,' he explains. 'It had to be enough to make it worth their while to bring it to me if they caught one, instead of taking it to the market – but not so much that the news got out that there was this crazy guy who was offering a fortune for a fish, resulting in an armada of fishing boats descending on Manado Tua, all on the hunt for a *rajah laut*. I didn't want to cause excess mortality. At first I was concerned that I would be inundated with coelacanths – which is why I didn't take the J. L. B. Smith route of plastering posters around entire villages.'

He soon realised he had been optimistic. The months passed and nothing came. Indonesia erupted into discord, the economy crashed, President Suharto resigned, and for a time, the streets of most big cities swarmed with angry and violent protesters. North Sulawesi, however, remained quiet. Mark went out on Om Lameh's boat at dusk to see the fishermen carefully let down their 100-metre net along the steep reef wall, attaching stones at its bottom to make sure it didn't drift – and at dawn to watch them pull it up, a dozen wiry men working synchronistically, like a professional tug-of-war team. Each time the net surfaced, Mark expected to see a coelacanth caught in the mesh. Each time he was disappointed. He studied the temperature and depth profiles of each fishing spot. It appeared that *rajah laut* lived at the same sort of depth and within the same temperature range (between sixteen and twenty degrees centigrade) as the Comoran coelacanth. In the middle of July, he raised the reward to Rp 1 million (£60 in the devalued Indonesian currency). 'I wished them good luck and told them to get on with it and bring me the fish.'

Twelve days later, and six thousand miles away, Marjorie Courtenay-Latimer, a sprightly ninety-one years old, was the guest of honour at a ceremony organised by the South African Mint to launch a limited, collector's edition of gold coelacanth coins. The evening was hosted by the Two Oceans Aquarium in Cape Town. Marjorie was introduced by Mike Bruton: 'Ladies and gentlemen, this evening we are in the presence of one of the great figures in South African and Western science,' he began. 'Someone who played a vital role in the greatest biological discovery of the century, the discovery of the first living coelacanth ... I refer of course, to Marjorie Courtenay-Latimer.'

Dressed in a neat black suit with a fake fur collar (which, she joked, was almost as old as the coelacanth), Marjorie approached the microphone to deliver a short speech. She wanted to thank the South African Mint, she said, for the great honour, and also to trace briefly the events which led to her discovery. As she spoke, about Captain Goosen, Bird Island and J. L. B. Smith, about seeing the beautiful blue fish for the first time and her absolute conviction that it had to be saved, she put down her prepared notes. Speaking only from her sharp memory, she transported the audience back sixty years, to a small museum in East London, and a young woman who was determined that the strange fish she had found was something special and had to be saved.

'I'll never stop on the coelacanth,' she said after the presentation. 'It's like perpetual motion. Sometimes I think I am sick to death of it, because there are so many other things I am interested in. I spent forty years at the museum and built it up from scratch, and it was about so much more than the coelacanth: my aim was to make it so that, if you only had fifteen minutes to spend, you would be able to enjoy it. I arranged all the fish with pools of water, corals and seaweed; all the birds and animals in natural settings. When I retired I went to live in Tsitsikamma for fifteen years, on a little farm, and wrote a book about the wild flowers in the national park down there. But I could never completely leave the coelacanth behind. On the fiftieth anniversary of its discovery, I was invited to the Comoros for a special ceremony at the museum. It was a wonderful experience: really I have been a very privileged person, a very spoilt person because of the coelacanth. The coelacanth put me on the world map, and I still get letters from all over – and lovely letters from schoolchildren – asking about the coelacanth. I try to reply to every one of them. And without the coelacanth, of course, I wouldn't have been here tonight to receive this lovely gold coin.'

At the same time as the director of the South African Mint was presenting Marjorie with a commemorative coelacanth coin made from the purest South African gold, across the Indian Ocean, six thousand miles away, a gold-flecked coelacanth emerged from its cave into the net Om Lameh had laid off the south-eastern coast of Manado Tua. A few hours later, on 30 July, Mark and Arnaz Erdmann were waiting at home for their boatman, Said, to arrive with the boat to take them to the city. 'He was running late, which is unusual for Said,' Mark recalls. 'At about 8.10 I saw the boat coming around the corner. There were several people on board, which was a bit unusual, but I just figured Said had offered them a ride into Manado. He came running up the steps, and leaned – rather too casually, I thought – in the doorway of my study. He had this huge grin on his face, but he was determined to act cool. Arnaz said, "Good morning. How are you?" at which point his cool broke. "We have a *rajah laut*!" he said.

'I looked down and I could see this large fish being held by Om Lameh's son in the shallows. We raced down the steps, Arnaz with video camera in hand, to the beach, very excited. I had a closer look, and saw that, yes, it was definitely a coelacanth – the real thing! What happened then was very much like JLB described. A million thoughts started racing around my head: What I should do, who I should tell?'

Marjorie Courtney-Latimer accepting her gold coelacanth coin, Cape Town 1998.

Arnaz filmed the coelacanth as it paddled slowly in a foot of water. It was clearly dying: it kept trying to roll over on to its back, and there was a tear in the delicate front dorsal fin. While she filmed Om Lameh, holding the coelacanth in his arms, Mark took photographs. At no point could he wipe the enormous smile off his face. He told Said and Om Lameh to pose the fish for the camera: tail up, fins out. But all the time, his mind was swirling with all he had to do.

'As the fish still looked alive – just – we decided to take it into deeper water for some more photographs. We grabbed our diving gear and my underwater camera, and took the coelacanth out to the reef flat, where we photographed it at a depth of two metres. However, as the visibility wasn't great, and it wasn't its natural habitat, I suggested we take it to the reef edge. Arnaz was against it: the fish was bleeding slightly and she was worried it might be grabbed by a shark. But I felt ready to do battle with any shark.'

Mark managed to convince her, and they towed the coelacanth behind the boat across the reef. As the water passed over its gills, it appeared to revive a little; it stopped trying to turn belly up and started to scull with its fins. They dived into the water with it, and guided it down a few more metres. It offered no resistance, and didn't try to struggle or escape.*

'We were down for about forty-five minutes. There was a strong current and bad visibility,' Mark recounts. 'I was very concerned about sharks, and kept looking around for them. I

*As they swam along the reef wall, a dive boat passed over them. Coincidentally, the man inside was Peter Scoones, the BBC cameraman who had filmed the dying coelacanth in the Comoros in 1977. This time, however, he was filming spawning clown fish, and remained oblivious to the presence of another coelacanth.

was also very caught up in the technical details of positioning it right for the camera, and trying to keep Arnaz out of the picture.'

'I swam beside the coelacanth with the rope in my hand,' Arnaz continues. 'Every now and then it tried to roll over, so I would support it and put it back into a swimming position. I was struck by its beauty – it looked like it was cloaked in golden armour. It swam in a very unhurried manner – it didn't even appear to be scared. Once in a while it would take a big gulp of water. It reminded me of a Spanish dancer, waving its fins like a flouncing skirt.'

'It was definitely feeling a bit better by that stage,' says Mark. 'When I moved in for some close-ups, I really had a chance to take it in. It was magnificent, each scale appeared to be flecked with gold. I touched it and it was very soft: I could put my arms around it and squeeze and it was more like holding a baby with soft, young flesh, than a big, hard fish. The thing that captivated me most was its eyes: they were large, and in certain lights were a luminescent, almost alien green, and they kept looking at me: wherever I went, its eyes followed. When we were in the shallows, the fisherman had said, "Look out for its mouth – don't let it bite you," but the thought never crossed my mind. It seemed very gentle and calm.'

They finished the film, then took the coelacanth back to the reef flat. Mark grabbed his dissecting kit, sample vials, liquid nitrogen container and alcohol, then they put the fish in some water in a big green cooler box and loaded it on to the boat. It was the largest chest they could find, but it was still too small for the coelacanth, which lay rather uncomfortably, its unusual tail poking out of the end, flapping its fins feebly from time to time.

Arnaz Mehta Erdmann swimming with the coelacanth.

'I was filled with excitement and adrenaline, but at the same time it was heartbreaking to see it slowly dying, especially having swum with it,' Mark recalls. 'It seemed so special, there was no hint of ferocity. At risk of slipping into anthropomorphisms, I had the impression of great gentleness and intelligence. I used to do a lot of spear fishing, and I have seen many dying fish – most of them are anything but dignified. They shake and thrash around. But the coelacanth seemed to me to be very stately. It was very very sad. I can honestly say that if it had looked more alive when we had been photographing it, I would have had the impulse to let it go.'

J. L. B. Smith would have appreciated Mark Erdmann's dilemma, yet he, too, would have made the same decision. Mark's swimming companion would, doubtlessly, have died within hours of its release – to this day, no coelacanth has survived the trauma of capture – and its precious inner secrets would have been swallowed in an instant by the predatory sharks which patrol the outer reef. Those secrets will be the start of a new chapter in the study of one of the world's most miraculous creatures, and may contribute to finding a way to ensure its continued existence for another 400 million years.

The coelacanth remained alive for most of the half-hour boat journey to Manado, moving less and less as time passed. Eventually, only its large green eyes showed any sign of life, and as the boat approached the harbour, they too became still, as quietly, with dignity, the Indonesian king of the sea died.

XII

TERRA INCOGNITA

On 24 September 1998, Mark Erdmann's article was published in *Nature*, as J. L. B. Smith's had been almost sixty years earlier. It was immediately dubbed 'the zoological sensation of the decade' and accorded an appropriate reception. Across the world, the coelacanth's picture was splashed over newspapers and on to television and the Internet. 'Second Home of Fish From Dinosaur Age is Found,' declared the headline in the *New York Times* above a long article which described the coelacanth as 'ugly but fascinating'. 'Fish on the slab was 360 million years old,' explained the *Daily Telegraph*, while the news channels – CNN, BBC, ABC and FOX – concentrated on the human story of the Erdmanns' honeymoon adventure. Jerry Hamlin's dinofish.com website received a record 6,500 hits. It was a far cry from Mr Adams of the *Daily Dispatch*, who had broken the story of *Latimeria chalumnae* in East London in 1939.

The coelacanth world, predictably, was thrown into a state of near hysteria. Certain scientists claimed to have been party to the secret all along, while others who had, forgot their initial scepticism and reached for their ocean current charts. Cyber lines between the major ichthyological centres buzzed as coelacanth specialists adjusted their assumptions, ate their words, and furiously debated the implications of the announcement.

Within minutes of the news reaching him, Hans Fricke was contemplating taking *Jago* to Manado Tua. His first feelings on hearing of a new population of coelacanths were mixed. 'It is great, fantastic news for science,' he said, with a big smile. 'It seems as if Old Fourlegs is tougher than we thought.' He was, however, struck by the possible effects: 'I feel sorry for the Comoran people. I hope it doesn't affect the conservation effort there. Maybe the coelacanth has a stronger chance of surviving in Indonesia, where they can learn from the mistakes made in the Comoros, and start trying to protect the fish from the beginning.'

For the majority of the scientists, however, the discovery of an Indonesian chapter was something to cheer about. Not only did it open up new areas of research, and provide them with fresh topics of discussion, but it unequivocally indicated a larger world coelacanth population than had been previously believed.

'It is just incredible that they were found in Indonesia, so far from where we had suspected the ancestral habitat to be in Madagascar or the Comoros,' said Robin Stobbs. 'It appears that both populations of coelacanths live at similar depths, at the same temperatures, in similar rocky and volcanic habitats, and together with the oil fish. Both are caught around the new moon (dark nights), and the deep shark nets used by Indonesian fishermen are identical to those used by ZeZe, the Malagasy fisherman who caught both the Madagascan specimens. The historical parallels, as well as the similarities in habitat, are remarkable. The Comoros and Madagascar were first settled by Indonesians nearly 2,000 years ago, who brought with them their fishing skills and techniques.'

More than 6,000 miles separate Manado from Moroni, comprising the vast belly of the Indian Ocean. Two millennia ago, strong men in longboats set sail from their Indonesian home. Riding the currents, they headed west into *terra incognita*. Eventually, the lucky ones landed on the wild and unpopulated islands off the African coast. Perhaps they found there, among the strange fauna and flora, a large and scaled, but familiar fish?

To Mark Erdmann's first glance, the Indonesian coelacanth appeared to be almost identical to its Comoran counterpart. The only noticeable difference was the colour: while live Comoran fish have always been described as steely blue with white splodges, the *rajah laut* was distinctly brown – with the same white markings, but also an intense glittering pattern of gold flecks covering its flanks. The golden shimmer, the prismatic effect of light reflected off the small, spiny denticles on its scales, had never been noted before, and could be taken as an indication, Mark believed, that his fish was not an exact replica of the Comoran *Latimeria chalumnae*. The proof would be determined by detailed genetic examination.

Barely an hour after his fish had died, Mark Erdmann had taken tissue samples of its major organs, and stored them in liquid nitrogen. Two months later – four days after the *Nature* article appeared – he was joined by Susan Jewett, a curator at the Smithsonian Institute in Washington DC, and fellow coelacanth enthusiast. She had come to Indonesia to help

Mark prepare and preserve his coelacanth. Together they took it from his landlady's freezer – its temporary home – carefully packed it into a polystyrene coffin, and escorted it by plane to Jakarta. From there, they took it by car to the Bogor Zoological Museum, an hour's drive away. It was rushed into the museum's sparkling dissection room, where an audience comprising the cream of Indonesia's scientific community had gathered. Mark and Susan donned long gauntlets and gas masks – protection against formalin fumes – as if ready to engage in biological warfare. They weighed (66 pounds/29.8 kilograms) and measured (49 inches/124 centimetres) the fish, and then injected it with preserving fluid, and arranged it in a lifelike position for exhibition. While they were preparing it, they found three small eggs, a clear indication that 'it' was a 'she'.

By this time, a rumour had emerged in coelacanth circles that Susan Jewett's express purpose was to smuggle the four-foot fish back to America in her suitcase. This, Mark was quick to deny. 'The first fish was always going to be Indonesia's,' he explained. 'Hopefully we will be able to arrange a CITES permit to send the next one to Washington, which is exactly half-way around the world from Jakarta.'

Susan Jewett returned to the Smithsonian to wait, her suitcase containing only the clothes she had left with. But by the time Mark returned to Manado, coelacanth mania had set in; the national media was drunk on the good news story, and the now-predictable collection of rumours and strange tales had started to emerge. A local man had been overheard in the fish market, offering Rp 2,000,000 (£120) for a *rajah laut*. An American scientist, desperate for some fresh coelacanth brain tissue, was apparently organising an expedition to the islands.

An Indonesian dive operator was insisting that *he* had found the coelacanth while diving, and a regular visitor to a guesthouse on Bunaken rather improbably insisted that the owner frequently cooks and serves coelacanth satay to his guests.

The French, inevitably, tried to trump Mark Erdmann's discovery. Within hours of the *Nature* announcement, a cryptozoological website was claiming that in 1995 a French fishery consultant called Georges Serres had caught a twenty-two-pound (ten-kilogram) coelacanth south of Java, while trying to fish for lobsters at night. He had salted and dried it, before giving it to the local fishery department, who promised to send it to the Oceanographic Institute in Jakarta. Inconveniently for Serres, all his belongings – including photographs of the fish, known locally as *ikan formar* – were stolen as he was about to leave the country, and the institute apparently has no record of having received the specimen. 'If we could find someone who could verify that the specimen prepared by Monsieur Serres was still at the institute in Jakarta, it would be a pretty "counter-scoop with a nice nationalistic flavour",' wrote one French scientist.

More worryingly, in the month following the announcement of the Indonesian coelacanth's whereabouts, the fishermen of Manado Tua – local superstars Om Lameh and Maxon and their crews – were approached by at least five groups of Japanese (or perhaps the same group five times), who offered them ever-increasing sums in an attempt to persuade them to co-operate in special coelacanth-hunting expeditions. Mark Erdmann is determined to prevent anyone from poaching a coelacanth from under his nose. Like so many people who have come into contact with it, he has fallen under the coelacanth's spell. Since 18 September 1997, when he saw his first fish in Manado fish market, he has thought and breathed of little else.

At the end of 1998, he made several trips to Jakarta to meet with the other members of a hastily convened coelacanth think-tank. The outcomes were upbeat; it was agreed that there would be a ministerial decree declaring the coelacanth to be part of Indonesia's national heritage, and as such to be protected for future generations. The proposal that there should be a coelacanth information centre was also greeted with enthusiasm.

In early 1999, the first results of the important DNA investigations came back from the laboratory. (The tests compared the DNA in the tiny mitochondria – thousands of bacteria-like power plants – in each cell, rather than the more complex DNA in the nucleus.) A sequence of 3,221 bases of the mitochondrial genetic blueprint of the Indonesian coelacanth was obtained, and after extensive massaging through a computer, it was calculated that they showed a 3.4 per cent divergence from the Comoran coelacanth's sequence. This is a substantial difference but not in itself necessarily enough, in Mark Erdmann's opinion, to call it a new species.* He wanted to wait for the results of detailed morphological comparisons between the Indonesian and the Comoran fish.

Others were not so cautious. In March 1999, just as the Erdmann team's results were awaiting publication in *Nature*, the French finally got their patriotic counter-scoop when a

---

*In the salamander world, such a divergence would indicate two separate species; while in bird and snail terms, it is well within the range of variation of a single species.

French catfish specialist, Laurent Pouyaud, working with a rival team from the Indonesian Institute of Sciences, reported the results of his own DNA investigations in *Comptes Rendus de L'Académie des Sciences*. From his results, Pouyaud appeared to have no hesitation in naming the Indonesian coelacanth as a new species, *Latimeria menadoensis* L. Pouyaud. Mark Erdmann was incensed at what he saw as a 'dishonourable act of scientific piracy' on the part of the Frenchman, and a mighty row broke out in the coelacanth world, of a magnitude only surpassed, perhaps, by the very public front page fight following J. L. B. Smith's 'theft' of the second coelacanth in 1952.

At a special meeting during the 1999 convention of the American Association of Ichthyologists and Herpetologists, a motion was passed to try to reverse Pouyaud's announcement, naming the fish, instead, after Arnaz Erdmann. However, this proved fruitless. Pouyaud had not contravened the laws of the current International Code of Zoological Nomenclature, and as he explained, his Indonesian colleagues had been eager for him to go ahead and trump the American team. His naming, thus, still stands, and the latest fracas was probably only to be expected, considering the incendary political history of the coelacanth.

Both scientific teams attempted to name the date when the two coelacanth 'families' parted ways. This was always going to be a 'best guess': as the Comoran *Latimeria chalumnae* has no close relatives, it is impossible accurately to determine when the genes split, but Erdmann's estimates put it between 5.5 million and 7.5 million years ago, while Pouyaud suggested a few million years later.

Either way, their results appear to answer the conundrum which has been exercising the minds of coelacanth junkies since the existence of the Indonesian fish was announced: which came first, *gombessa* or *rajah laut*? Seven and a half million years ago, there were no Comoro Islands, the volcanoes had yet to erupt out of the sea to create land. Great tectonic movements were still changing the face of the earth. It is possible, Mark Erdmann suggested, that the massive plate movements which led to the formation of the Indo-Australian Arc and the resultant separation of the Indian and Pacific Oceans in the Miocene period (from 25 million years ago), cut a part of the coelacanth population off from the rest. If this is what happened, it is conceivable that there are more coelacanths, in different locations across the globe. 'It seems highly unlikely that the living coelacanth exists only in two small, highly disjunct populations,' Mark Erdmann wrote in *Nature*. Fired by the enthusiasm of the response, he was eager to start showing his photographs in coastal villages and fish markets around the Indonesian archipelago.

The possibility that there might be coelacanths off other exotic islands will undoubtedly be spur enough for a whole new generation of adventurers and enthusiasts to print up their own reward posters and start searching the more remote fringes of the Indian Ocean. The suggestion that the two silver coelacanth models might have been of Philippine rather than Meso-American origin, and carried back from the Philippines (only a few hundred miles from Manado) by Spanish merchant traders to Toledo and Bilbao, adds fuel to their fires.

The rumours surrounding the coelacanth will doubtlessly continue to attract enthusiasts and eccentrics, or perhaps to drive its admirers to eccentricity. A few years ago, the daughter of a German count sent Hans Fricke a beautifully handwritten treatise the length of a book, extolling her theory of how and why the silver coelacanths were made. She had acquired this knowledge, she explained, from free-floating electro-magnetic rays. An East London taxi driver, C. D. Harrald, has spent his entire life telling people that he was the man who nearly refused to take Marjorie Courtenay-Latimer, her assistant Enoch and 'that stinking fish' from the docks to the museum on 22 December 1938. What he fails to recall, however, was that he was too young to have had a licence at that time, and that it was his uncle – who died in 1969 – who was in fact driving the taxi.

Even when the coelacanth was known only as a fossil, it captured the imagination, and it appears to have maintained its allure. In the months after the existence of an Indonesian population became public knowledge, numerous scientific and photographic expeditions were being planned, and in late 1999, Mark Erdmann joined Hans Fricke and his team of 'Jagonauts' on an expedition to try to find and film a live coelacanth in its natural, Indonesian environment. After weeks of fruitless searching, just as they were preparing to pack up and go home empty-handed, they found two coelacanths, 360km southwest of Manado Tua.

'It seemed like an almost impossible task to find the beasts in these tremendously fast, turbid waters,' recounted Hans Fricke, days after his return. 'Sometimes it was really risky, but it was worth it. We shot a wonderfully dramatic film and saw absolutely amazing creatures in the deep water.' 'It was an amazing experience,' said Mark Erdmann. 'I bet J. L. B. Smith would have loved to have seen it.'

However, Fricke was not convinced that Sulawesi was the coelacanth's native home. 'We cannot exclude the possibility that the coelacanths of north Sulawesi are derived from a different area and drifted there with oceanic currents,' he reported in *Nature* on 6 January 2000.

> *The dominant current . . . is the southerly Mindanao current, indicating that the potential source population may be in the southern Philippines or remote Pacific islands . . . The biogeography of the new coelacanth population remains enigmatic, although perhaps this is for the best. An undiscovered home is probably the best possible protection for these endangered fish.*

This, sadly, is a forlorn hope. Where there is a glimmer of a coelacanth, adventurers quickly follow. Without doubt, more coelacanths will be found in Indonesia, more photographs will be taken and films made, and in time, the hidden world of *rajah laut*, swimming below the reefs of Manado Tua, will be revealed to us. The preserved Indonesian coelacanths will sit alongside their Comoran cousins in the museums of the world. With any luck, many will continue to evade our nets and lines. 'The fact that living coelacanths could escape detection in an area well studied by ichthyologists for over 100 years is wonderful,' Mark enthused. 'It is a humbling and exciting reminder that humans have by no means conquered the oceans, and provides hope that "Old Fourlegs" is more abundant and resilient than we initially dared hope.' But, as Marjorie Courtenay-Latimer, who was responsible for first bringing this magnificent creature to our attention over sixty years ago, said recently, 'Perhaps now, it is time to leave the coelacanth in peace.'

Human beings must appear mere parvenus in the ledgers of coelacanth history. Our important events – from learning to make stone tools, to walking on the moon – are but a small bleep to the silent witness at the bottom of what we presume to call 'our' oceans. It is comforting to imagine the coelacanth swimming quietly around, unaware of the crazy things that take place, surviving far greater holocausts than those we have known, and continuing to exist after this extraordinary duration of time.

SELECTED READING

I am indebted to a large number of books and articles, only some of which I am able to acknowledge:

BOOKS

Anthony, Jean: *Opération Coelacanthe* (Arthaud, 1976)

Barnett, Peter: *Sea Safari with Professor Smith* (South African Association for Marine Biological Research)

Broad, William: *The Universe Below* (Simon & Schuster, 1997)

Forey, Peter: *History of the Coelacanth Fishes* (Chapman & Hall, 1998)

Forte, Richard: *Life: An Unauthorised Biography* (HarperCollins, 1997)

Ley, Willy: *Exotic Zoology (Viking Press, 1959)*

Long, John A.: *The Rise of Fishes – 500 Million Years of Evolution*, (University of New South Wales Press, *1995*)

Millot, Jacques: *Le Troisième Coelacanthe* (Le Naturaliste Malgache, 1955)

Millot, Jacques and Anthony, Jean: *L'anatomie de Latimeria chalumnae* (Centre Nationale de Réchèrches Scientifiques, 1960–78)

Smith, J. L. B.: *Sea Fishes of Southern Africa* (Central News Agency, 1949)

    Old Fourlegs: The Story of the Coelacanth (Longman, Green, 1956)

Thomson, Keith: *Living Fossil* (Norton, 1991)

Ward, Peter Douglas: *On Methuselah's Trail* (W. H. Freeman, 1991)

PAMPHLETS

J. L. B. Smith: His Life, Work, Bibliography and List of New Species (M. M. Smith, Rhodes University, 1969)

The Life and Work of Margaret M. Smith (J. L. B. Smith Institute of Ichthyology, undated)

Ichthos: Tribute to Margaret Smith (J. L. B. Smith Institute of Ichthyology, 1987)

Ichthos: The Coelacanth Jubilee (J. L. B. Smith Institute of Ichthyology, 1988)

Ichthos: J. L. B. Smith Commemorative Edition (J. L. B. Smith Institute of Ichthyology, 1997)

ARTICLES

Balon, E.; Bruton, M. and Fricke, H.: 'A fiftieth anniversary reflection of the living coelacanth' (*Environmental Biology of Fishes*, 1988)

Bergh, W.; Smith, W.; Botha, W. and Laing, M.: 'The place of Natal Command in the history of world science' (Spectrum, 1992)

Bruton, M.: 'The living coelacanth fifty years later' (*Transactions of the Royal Society of South Africa*, 1989)
'The coelacanth – can we save it from extinction?' (*World Wildlife Fund Reports*, 1989)
'The mingled destinies of coelacanths and men' (*Ichthos*, 1992)

Bruton, M.; Cabral, Q. and Fricke, H.: 'First capture of a coelacanth off Mozambique' (*South African Journal of Science*, 1992)

Conant, E. B.: 'An historical overview of the literature of Dipnoi' (*Journal of Morphology*, 1986)

Courtenay-Latimer, E.: 'Diaries' (unpublished, courtesy of Dr M. Courtenay-Latimer)

Courtenay-Latimer, M.: 'My story of the first coelacanth' (*Occidental Papers of the California Academy of Science*, 1979)
'Reminiscences of the discovery of the coelacanth' (*Cryptozoology*, 1989)

De Sylva, D.: 'Mystery of the silver coelacanth' (*Sea Frontiers*, 1966)

Dugan, J.: 'The fish' (*Colliers*, 1955)

Erdmann, M.; Caldwell, R. and Moosa, K.: 'An Indonesian "King of the sea"' (*Nature*, 1998)

Forey, P.: 'Golden jubilee for the coelacanth' (*Nature*, 1988)
'Blood lines of the coelacanth' (*Nature*, 1991)

Fricke, H.: 'The fish that time forgot' (*National Geographic*, 1988)
'Im Reich der lebenden Fossilien' (*Geo*, 1987)
'Living coelacanth: values, eco-ethics and human responsi-

bility' (*Marine Ecology Progress Series*, 1997)

Fricke, H. and Hissmann, K.: 'Natural habitat of the coelacanth' (*Nature*, 1990)

Fricke, H. and Plante, R.: 'Habitat requirements of the living coelacanth' (*Naturwissenschaften*, 1988)

Greenwood, P. H.: 'Fifty years a "living fossil"' (*Biologist*, 1989)
'Latimeria chalumnae – the living coelacanth' (*Ichthos*, 1993)

Hall, M.: 'The survivor' (*Harvard Magazine*, 1989)

Heemstra, P. and Compagno, L.: 'Uterine cannibalism and placental viviparity in the coelacanth? A skeptical view' (*South African Journal of Science*, 1989)

Heemstra, P., Freeman, A., Yan Wong, H., Hensley, D. and Rabesabdratana, H.: 'First authentic capture of a coelacanth off Madagascar' (*South African Journal of Science*, 1996)

Hissmann, K.; Fricke, H. and Schauer, J.: 'Population monitoring of the coelacanth' (*Conservation Biology*, 1998)

Hissmann, K. and Schauer, J.: 'Fossil hunt' (*Diver*, 1991)

Millot, J.: 'Notre Coelacanthe' (*Revue Madagascar*, 1953)
'First observations on a living coelacanth' (*Nature*, 1955)

Morris, E. and A.: 'In pursuit of the coelacanth' (*Pacific Discovery*, 1973)

Munnion, C.: 'Remembering old fourlegs' (*Optima*, 1988)

Plante, R.; Fricke, H. and Hissmann, K.: 'Coelacanth population, conservation and fishery activity at Grande Comore' (*Marine Ecology Progress Series, 1998*)

Schauer, J.: 'The privacy of a living fossil' (*Underwater*, 1992)

Smith, J. L. B.: 'A living fish of the Mesozoic type' (*Nature*, 1939)
'A surviving fish of the order Actinistia' (*Transactions of the Royal Society of South Africa*, 1939)
'A living coelacanth fish from South Africa' (*Transactions of*

the *Royal Society of South Africa*, 1940)

'The second coelacanth' (*Nature*, 1953)

Smith, M.: 'The search for the world's oldest fish' (*Oceans*, 1970)

Stobbs, R.: 'The coelacanth enigma' (*The Phoenix*, 1989)

'The Comoro Islands' traditional artisanal fishery' (*Ichthos*, 1990)

'Hiriako – The broken thread' (*Ichthos*, 1996)

'Eric Ernest Hunt – the aquarist' (*Ichthos*, 1996)

'Gone Fishin' – for a purgative' (*Ichthos*, 1998)

Vicente, N.: 'Un coelacanth à Madagascar' (*Oceanorama*, 1997)

White, E. I.: 'One of the most amazing events in the realm of natural history in the twentieth century' (*London Illustrated News*, 1939)

INDEX

Figures in italics refer to illustrations.

All Fourth Estate books are available from your local bookshop, or can be ordered direct (FREE UK p&p) from:

Fourth Estate, Book Service By Post, PO Box 29, Douglas, I-O-M, IM99 1BQ

Credit cards accepted.

Tel: 01624 836000   Fax: 01624 670923

Or visit the Fourth Estate website at:
www.4thestate.co.uk

*Please state when ordering if you do **not** wish to receive further information about Fourth Estate titles.*